FOREWORD

In the future, nuclear fuel will reach high burn-up, and recycling of spent fuel will possibly have to be performed after shorter cooling times to be economically and ecologically attractive. Dry reprocessing using pyrochemical methods has received the attention of some research institutes due to its potential advantages of compactness in the plant design, criticality safety considerations as well as stability against high radiation dose.

Recently, partitioning and transmutation (P&T) of long-lived fission products and minor actinides has been considered as one of the future options of waste management. In order to achieve effective transmutation, multi-recycling of fuel with much higher burn-up and shorter cooling time will be required. Therefore, the role of pyrochemical reprocessing becomes more important for the implementation of the P&T option.

The proposed processes are complex and necessitate the use of highly controlled atmospheres to avoid hydrolysis and precipitation reactions. Except for the pilot-scale demonstration of the pyro-electrolysis at Argonne National Laboratory (Idaho) and the Russian Institute of Atomic Reactors, all other studies are at the laboratory scale. Much R&D work is required in order to upgrade the pyrochemical separation process to the current level of industrial aqueous reprocessing.

For this reason, the OECD/NEA Nuclear Science Committee decided to organise a workshop to exchange information about recent activities on pyrochemical separations and to make a recommendation about the future activities of the OECD/NEA in the field.

The Workshop on Pyrochemical Separations, co-organised by the European Commission and hosted by CEA Valrho, was successfully held in Avignon, France, on 14-15 March 2000, with more than 70 participants from 14 countries and 3 international organisations. There were 21 oral presentations and 8 posters on the following topics:

- National and international R&D programmes on pyrochemical separations.

- Role and requirements of pyrochemical reprocessing in future fuel cycles.

- Recent progress in research activities, including:

    - Basic data (thermodynamics, reaction mechanisms, reaction kinetics, molecular modelling).

    - Results and experiences from process experiments.

    - Process simulation and process design.

It was recommended that an expert group be set up to prepare a state-of-the-art report on pyrochemical separations. The report, furnishing timely and coherent coverage of situations in each country and providing a common scientific base on pyrochemical separations, will be very useful for promoting efficient international collaboration.

OECD PROCEEDINGS

Proceedings of the Workshop on

PYROCHEMICAL SEPARATIONS

Co-organised by the
European Commission

Hosted by
Commissariat à l'Énergie Atomique Valrhô

Avignon, France
14-16 March 2000

NUCLEAR ENERGY AGENCY
ORGANISATION FOR ECONOMIC CO-OPERATION AND DEVELOPMENT

ORGANISATION FOR ECONOMIC CO-OPERATION AND DEVELOPMENT

Pursuant to Article 1 of the Convention signed in Paris on 14th December 1960, and which came into force on 30th September 1961, the Organisation for Economic Co-operation and Development (OECD) shall promote policies designed:

- to achieve the highest sustainable economic growth and employment and a rising standard of living in Member countries, while maintaining financial stability, and thus to contribute to the development of the world economy;
- to contribute to sound economic expansion in Member as well as non-member countries in the process of economic development; and
- to contribute to the expansion of world trade on a multilateral, non-discriminatory basis in accordance with international obligations.

The original Member countries of the OECD are Austria, Belgium, Canada, Denmark, France, Germany, Greece, Iceland, Ireland, Italy, Luxembourg, the Netherlands, Norway, Portugal, Spain, Sweden, Switzerland, Turkey, the United Kingdom and the United States. The following countries became Members subsequently through accession at the dates indicated hereafter: Japan (28th April 1964), Finland (28th January 1969), Australia (7th June 1971), New Zealand (29th May 1973), Mexico (18th May 1994), the Czech Republic (21st December 1995), Hungary (7th May 1996), Poland (22nd November 1996), Korea (12th December 1996) and the Slovak Republic (14th December 2000). The Commission of the European Communities takes part in the work of the OECD (Article 13 of the OECD Convention).

NUCLEAR ENERGY AGENCY

The OECD Nuclear Energy Agency (NEA) was established on 1st February 1958 under the name of the OEEC European Nuclear Energy Agency. It received its present designation on 20th April 1972, when Japan became its first non-European full Member. NEA membership today consists of 27 OECD Member countries: Australia, Austria, Belgium, Canada, Czech Republic, Denmark, Finland, France, Germany, Greece, Hungary, Iceland, Ireland, Italy, Japan, Luxembourg, Mexico, the Netherlands, Norway, Portugal, Republic of Korea, Spain, Sweden, Switzerland, Turkey, the United Kingdom and the United States. The Commission of the European Communities also takes part in the work of the Agency.

The mission of the NEA is:

- to assist its Member countries in maintaining and further developing, through international co-operation, the scientific, technological and legal bases required for a safe, environmentally friendly and economical use of nuclear energy for peaceful purposes, as well as
- to provide authoritative assessments and to forge common understandings on key issues, as input to government decisions on nuclear energy policy and to broader OECD policy analyses in areas such as energy and sustainable development.

Specific areas of competence of the NEA include safety and regulation of nuclear activities, radioactive waste management, radiological protection, nuclear science, economic and technical analyses of the nuclear fuel cycle, nuclear law and liability, and public information. The NEA Data Bank provides nuclear data and computer program services for participating countries.

In these and related tasks, the NEA works in close collaboration with the International Atomic Energy Agency in Vienna, with which it has a Co-operation Agreement, as well as with other international organisations in the nuclear field.

TABLE OF CONTENTS

EXECUTIVE SUMMARY

The Workshop on Pyrochemical Separations, co-organised by the European Commission and hosted by CEA Valrho, was successfully held in Avignon, France on 14-15 March 2000 with more than 70 participants from 14 countries and 3 international organisations. There were 21 oral presentations and 8 posters in the following sessions:

- Session I: National and international R&D programmes on pyrochemical separations.

- Session II: Role and requirements of pyrochemical reprocessing in future fuel cycles.

- Session III: Recent progress in research activities:

  a. Basic data (thermodynamics, reaction mechanisms, reaction kinetics, molecular modelling, etc.).

  b. Results and experiences from process experiments.

  c. Process simulations and process design.

- Session IV: Possible international collaboration (open discussion).

In Session I, there were five presentations covering the programmes carried out by the European Commission (Fifth Framework Programme), the USA (ATW), Russia (SSC RIAR), the UK (BNFL) and France (CEA).

In Session II, the role and requirements of pyrochemical separations with regard to technological and ecological advantages as well as economic and social (proliferation) aspects were addressed. Several separation system concepts were presented.

Session III, Part A, was dedicated to the basic data of pyrochemical separations. New experimental data obtained through intensive international research co-operation were reviewed and an ongoing effort for the molten salt database were presented. Stability (potential-pO^{2-} diagram) of two oxidation states of plutonium (0 and III), redox reaction parameters obtained by transient electrochemical techniques and the relation between thermodynamic stabilities and extractabilities were reported.

In Session III, Part B, presentations were made regarding results and experiences from process experiments, such as the electrometallurgical technology (EMT) at Argonne National Laboratory, nitride/pyroprocess for minor actinide transmutation, reduction of unirradiated transuranic oxides and pyrometallurgical processing for metal fuel and high-level waste.

Session III, Part C was related to process simulation and design. MoNiCr alloy was tested as a structural material. The extraction mechanism and the system performance were discussed for further development of the extraction system. Three-dimensional reaction diagrams (potential-acidity-temperature) for conceiving the skeletal structure of a pyrochemical process were reported.

In the poster session, eight posters were presented covering the following topics: speciation of Tc by means of EXAFES and NMR; feasibility of molten salt reactor technology for treatment of Pu, minor actinides and fission products; recovery of noble metals and tellurium from fission products; the study of uranium oxidation for pyrochemical reprocessing; Am separation from rare earth by electrorefining; pyro-oxidation of Pu and Am and vacuum distillation; an electrochemical study on Eu(III)/Eu(II) in NaCl-KCl melt; and an attempt using the generalised perturbation theory to predict fuel composition in the molten salt reactor.

In Session IV, Professor Madic, the workshop chairman, briefly introduced a proposal for setting up a task force under the auspices of the NEA Nuclear Science Committee which would be responsible for the preparation of a state-of-the-art report on pyrochemical separations (see Annex 3). The floor was then open for discussion.

The participants expressed strong interest in the proposal, as the report would preserve the knowledge previously obtained and would provide a common scientific base for reviewing the research programmes in each country.

The following comments were made on the scope and objectives of the proposed state-of-the-art report. The report would: examine possible nuclear fuel cycle scenarios within the next 50 years; include applications, advantages and disadvantages of pyrochemical separations; cover fission product separation and industrial work; and provide a time plan for future R&D.

The chairman concluded from the above discussions that a large majority of the audience was very supportive of establishing an NEA task force for preparing a state-of-the-art report on pyrochemical separations. It was agreed by the participants that the next workshop on pyrochemical separations would be held after completion of the report.

OPENING SESSION

WELCOME ADDRESS

Charles Madic
CEA/DCC, Saclay, France
Chairman of the Workshop

In recent years, a renewed interest by the media in fused salts for new nuclear fuel cycles or new nuclear waste management strategies has been observed. That is why, at the annual meeting of the Nuclear Science Committee in June 1999, I made a proposition to organise an international workshop on pyrochemical separations as a preliminary step before the preparation of an OECD/NEA/NSC state-of-the-art report on the subject. This proposition was accepted by the NSC, and soon after it was decided that the European Commission would also be involved in the organisation of the workshop.

As the CEA launched a programme in the field of pyrochemical separations in 1999 and created a dedicated laboratory in ATALANTE at Marcoule, I proposed to organise this workshop at the Palais des Papes in Avignon, near Marcoule. Dr. Philippe Brossard, from Marcoule, kindly accepted to be the principal local organiser of the workshop. I would like to thank him and his team warmly for the good work they did to organise the workshop.

Marcoule is a place where a large expertise has been gained over the last thirty years in the field of "pyroprocessing", with the highly successful development of the vitrification process for highly active nuclear wastes, which was first operated in the world at the industrial level at Marcoule, in the so-called AVM plant, which was hot commissioned in 1978. In the coming years, Marcoule will also be the place where pyrochemical separation processes will be developed in the framework of the CEA's programme related to the law of 30 December 1991 concerning the development of new nuclear waste management strategies.

On behalf of Dr. N. Camarcat, Director of the Fuel Cycle, and Mr. J.Y. Guillamot, Director of the CEA's Valrhô, I am pleased to welcome you at the Palais des Papes in Avignon.

I wish you a very interesting workshop and a pleasant stay in Avignon.

OPENING ADDRESS

Philippe Savelli
Deputy Director
Science, Computing and Development
OECD Nuclear Energy Agency

Ladies and gentlemen, it is a great pleasure for me to welcome you to this workshop on behalf of the OECD Nuclear Energy Agency.

For those of you who are not familiar with the Nuclear Energy Agency, the NEA is a subsidiary body of an international organisation, the OECD, based in Paris. Its membership resembles closely that of the OECD and is comprised of 27 of the most developed countries in the world. The basic mission of the NEA is to assist its Member countries in maintaining and further developing, through international co-operation, the scientific, technological and legal bases required for a safe, environmentally friendly and economical use of nuclear energy for peaceful purposes. Its role is also to provide authoritative assessments and to forge common understandings on key issues, as input to government decisions on nuclear energy policy. The Agency pursues this mission, for example by pooling Member countries' expertise in co-operative projects, by disseminating important information, by developing consensus opinions and by arranging meetings and workshops such as this one.

The NEA programme of work addresses all the key issues in the nuclear energy area, such as nuclear safety, radioactive waste management, radiation protection, legal aspects, economics and nuclear science. In addition, one sector of the NEA, the NEA Data Bank, gives a direct cost-free service on nuclear data and computer programs to scientists in Member countries.

Traditionally, a large part of the NEA's scientific programme of work in science has been devoted to reactor and fuel cycle physics. However, in recent years, this programme has extended to also cover nuclear fuel chemistry issues. One preoccupation in the nuclear fuel cycle is the separation of radioactive elements from other materials. For this reason, the NEA established a small group of experts to review the different techniques and chemical processes used in the separation of actinides. The group produced a state-of-the-art report entitled *Actinide Separation Chemistry in Nuclear Waste Streams and Materials*. The report was published in 1997.

As a follow-up to this state-of-the-art report, the NEA arranged the Workshop on Long-Lived Radionuclide Chemistry in Nuclear Waste Treatment. This workshop, held on 18-20 June 1997 at Villeneuve-lès-Avignon, France was organised by Professor Charles Madic. The objective of this workshop was to provide up-to-date information on the chemistry of radionuclides and to provide guidance on future activities that could be undertaken in the framework of the NEA. The workshop identified several topics which deserve special attention and the outcome was a recommendation that the NEA organise two additional workshops:

- The Workshop on Speciation, Techniques and Facilities for Characterisation of Radioactive Materials at Synchrotron Light Source, which was held at Grenoble on 4-6 October 1998 (the proceedings have been published). A second meeting on the same topic will be held in Grenoble next September.

- The Workshop on Evaluation of Speciation Technology, which was held in Japan on 26-28 October 1999 (publication of the proceedings is pending).

Recently, much interest has been shown in the pyrochemical process because of its ability to treat high burn-up fuel and to shorten the cooling time of this fuel. In addition its adaptability to multi-recycle the targets in transmutation systems is recognised, as Professor Madic mentioned in his speech. However, much R&D work will be required in order to upgrade the pyrochemical separation process to the level of present industrial aqueous reprocessing.

For this reason, the OECD/NEA Nuclear Science Committee decided last June to produce a state-of-the-art report on pyrochemical processes and to organise this workshop to prepare this report by exchanging information and making recommendations about possible international co-operation in the field.

Concerning the organisation of this workshop, I would first of all like to thank Charles Madic for the crucial role he has played in the arrangements, both as initiator and also as a chairman of the workshop. I know that organising a workshop on this scale is very hard work, and I especially wish to thank Dr. Philippe Brossard and his team at CEA Valrhô for hosting it. I am also grateful to the Scientific Advisory Board for all their support in making the workshop successful and stimulating, and to the European Commission for its role in the co-organisation.

Thank you all for being here. I wish you an interesting, instructive and profitable three days.

SESSION I

National and International R&D Programmes on Pyrochemical Separations

Chairs: G. Choppin, B. Boullis

RESEARCH WORK ON RADIONUCLIDE PARTITIONING FROM NUCLEAR WASTE CO-FINANCED BY THE EUROPEAN COMMISSION

Michel Hugon
European Commission
200, rue de la Loi, B-1049 Brussels

Abstract

There has been a renewal of interest for partitioning and transmutation of radioactive waste since the end of the 1980s, more especially in Japan and in France. More recently, the concept of accelerator-driven systems for radioactive waste incineration has been gaining momentum in Europe, the USA and Japan. However, a prerequisite for efficient transmutation of the critical long-lived radionuclides is the development and the demonstration of processes for their selective separation (partitioning) from high level liquid waste. The two main processes are hydro-metallurgical and pyrochemical processes.

Research work on hydro-metallurgical processes only has been partly financially supported by the European Commission in shared-cost actions in different European laboratories during the 3rd and the 4th Framework Programmes (1990-1999) and has been part of the institutional programme of the Institute of Transuranium Elements in Karlsruhe.

This paper gives a brief overview of the ongoing and future research projects on radionuclide partitioning, partly financed by the European Commission in its different research programmes. At present, two projects are still running in the framework of the INCO-COPERNICUS specific programme under the 4th Framework Programme. INCO-COPERNICUS was aiming at developing international co-operation between the EU and third countries and in this case more especially with the countries of Central Europe and the New Independent States (NIS). The first project is dealing with the synthesis of dicarbollide derivatives and their test on radioactive waste in collaboration with Czech research institutes. The second one concerns the development and testing of new organophosphorous ionophores for the decontamination of radioactive waste and involves Russian and Ukrainian laboratories.

Three research projects have been selected for funding in the key action on nuclear fission of the 5th Framework Programme. Two of them are related to hydro-metallurgical processes for the selective extraction of actinides from high level waste, the first one with nitrogen polydendate ligands or acidic sulphur-bearing ligands and the second one with functionalised calixarenes. The third project will assess pyrometallurgical flowsheets to process spent fuel and targets and in particular the separation of actinides from lanthanides by either salt/metal extraction or electrorefining. This project will also involve Czech and Japanese research institutes.

A sixth project might be funded by the International Science and Technology Centre (ISTC), which is an intergovernmental organisation with the following parties: EU, Japan, USA, Norway, Republic of Korea and Russian Federation. The ISTC finances and monitors science and technology projects to ensure that the NIS scientists, especially those with expertise in developing weapons of mass destruction, are offered the opportunity to use their skills in the civilian fields. The research project is proposed by Russian research institutes and is aiming at re-evaluating the molten salt nuclear fuel technology potential for the development of a molten salt burner concept.

Introduction

Research work on aqueous processes for partitioning has been partly financed by the EU in shared-cost actions during FP3 and FP4 (1990-1999). Work has also proceeded as part of institutional programmes of JRC/ITU in Karlsruhe.

At present, two projects still exist under FP4 in the INCO-COPERNICUS programme (international co-operation between EU and CCE, NIS). Additionally, three research projects (one on pyrochemistry) have been selected for funding in key action on nuclear fission in FP5. These projects will probably start in July 2000, and will last for a duration of 36 months.

Also, one proposal on molten salt technology might be funded by EU and other parties in the framework of ISTC (EU, J, USA, N, Rep. K, RF).

Selective separation of M^{2+} and M^{3+} radionuclides, namely of Sr and actinides, from nuclear waste by means of chelating hydrophobic cluster anions

The following institutions are participating in this programme: Barcelona University, Katchem Praha, CEN Cadarache, Inst. of Inorganic Chemistry Řež, Lyon University, Nuclear Res. Inst. Řež. The project began on 01/02/99, and has a projected end date of 31/01/01.

Work programme

Two tasks have been set for this project. Firstly, from the bis-dicarbollide cobaltate (-) anion patented in Czechoslovakia, it will be attempted to synthesise new extractants for the selective removal of strontium and actinides from medium and high level liquid waste. Secondly, the new extractants will be tested on simulated and/or real waste.

Development of technologies on efficient decontamination of radioactive wastes based on new organophosphorous ionophores

The following institutions will be pursuing this goal: Strasbourg University, ENEA Saluggia, Khlopin Institute St. Petersburg, Institute of Inorganic Chemistry Novosibirsk, Institute of Organic Chemistry Kiev. This project began on 01/11/98, and has a projected end date of 31/10/00.

Work programme

The work programme of this project consists of:

- Synthesis of new organophosphorous ionophores (functionalised crown-ethers and calixarenes).

- Development of solid extractants and sorbents.

- Semi-industrial tests of technologies concerning the extraction and safe storage of hazardous radionuclides.

- Physico-chemical and molecular modelling studies to provide guidance for the design of efficient extractants.

Selective extraction of minor actinides from high activity liquid waste by organised matrices (CALIXPART)

Participants include: CEA, ECPM Strasbourg, University of Mainz, Micromod, University of Parma, University of Twente, University of Strasbourg, University of Liège, University of Madrid and CIEMAT Madrid.

The objective of this activity is the design and implementation of new molecules able to complex with high efficiency and selectivity minor actinides in order to remove them in one step from HLW.

Work programme

In this project, the participants will undertake to:

- Synthesise ligands for cations using macrocycles such as calixarenes, resorcinarenes and cyclotriveratrylenes.

- Test the extraction capability of the new compounds.

- Determine the structure of the complexes by X-ray diffraction and NMR spectroscopy measurements.

- These measurements provide an input to molecular modelling, which guides the design of further improved compounds.

- Test the radiochemical stability of the new ligands and test their extraction capability with real HLW.

Partitioning: New solvent extraction processes for minor actinides (PARTNEW)

Participants of this project include: CEA, University of Reading, Chalmers University, ITU Karlsruhe, ENEA Saluggia, Poli. Milano, FZK Karlsruhe, FZJ Jülich, CIEMAT Madrid and the University of Madrid.

The objective is to design solvent extraction processes of Am[III] and Cm[III] from genuine high level raffinates (HAR) or concentrates (HAC) issuing PUREX reprocessing.

Work programme

The work programme includes:

- Studies of DIAMEX processes for co-extraction of Am[III] and Cm[III] from acidic HAR or HAC.

- Studies of processes for Am[III]/Cm[III] separation.

- Studies of SANEX processes for Am[III]/Cm[III] group separation from trivalent lanthanides from acidic feeds (or directly from HAR or HAC):

  - With polydendate nitrogen ligands.

  - With dithiophosphinic acid synergistic mixtures.

  - With new S, Se or Te-bearing ligands.

Pyrometallurgical processing research programme (PYROREP)

Participants include: CEA Marcoule, CIEMAT Madrid, ENEA Casaccia, ITU Karlsruhe, NRI Řež, BNFL Sellafield, AEA Harwell and CRIEPI.

The objective of this activity is to assess pyrometallurgical processing for the separation of U, Pu and minor actinides from spent fuel, metal target or reduced oxide feed.

Work programme

The following items are on the work programme:

- Separation of actinides and lanthanides by:

  - Molten fluoride salt/metal extraction.

  - Electrorefining in molten fluoride or chloride with liquid or solid cathodes.

- Salt decontamination from actinide traces by:

  - Molten fluoride salt/metal extraction.

  - Electrolytic method in a chloride medium.

- Selection and testing of materials compatible with molten fluoride salt media at high temperatures.

- Waste studies: leaching tests on sodalites.

Molten salt technology for radioactive waste and plutonium treatment (ISTC # 1606)

Participants include: VNIITF Chelyabinsk, Kurchatov Institute Moscow, IVTEH Sverdlovsk and VNIIHT Moscow.

The objective of this study is to reassess the potential of molten salt technology for the safe, low waste producing and proliferation resistant treatment of radioactive waste and plutonium management.

Work programme

The work programme includes:

- Molten salt burner conceptual system development (core design including thermal, epithermal and fast systems, fuel handling and process flowsheets, waste management, safety, economics, licensing, non-proliferation).

- Measurement of basic properties of molten salt fuel compositions, including experiments with plutonium and stability under irradiation.

- Test of key structural materials (modified hastelloy-NM for primary circuit, graphite as reflector/moderator, materials with Mo and W coating for fuel clean-up system).

- Preliminary evaluation of the cost of a molten salt plant.

Conclusion

For the past ten years, the EU has been partially funding research work on radionuclide partitioning from HLW by aqueous processes, and continues its support for this field during FP5.

Currently, there is a renewal of interest in the EU for pyrochemical processes (to recycle higher burn-up spent fuel after shorter cooling times in the future).The EU's continued interest is evidenced by the decision to select the PYROREP project for funding in FP5, as well as by the support offered for the ISTC# 1606 Russian proposal on molten salt.

The EU supports the OECD/NEA proposal on pyrochemical reprocessing, which it believes will strengthen the co-operation links with Japan, Republic of Korea and the USA in this field.

PYROCHEMICAL SEPARATIONS TECHNOLOGIES ENVISIONED FOR THE US ACCELERATOR TRANSMUTATION OF WASTE SYSTEM

James J. Laidler
Chemical Technology Division
Argonne National Laboratory
Argonne, IL 60439-4837
United States of America

Abstract

A programme has been initiated for the purpose of developing the chemical separations technologies necessary to support a large accelerator transmutation of waste (ATW) system capable of dealing with the projected inventory of spent fuel from the commercial nuclear power stations in the United States. The baseline process selected combines aqueous and pyrochemical processes to enable the efficient separation of uranium, technetium, iodine and the transuranic elements from LWR spent fuel. The diversity of processing methods was chosen for both technical and economic factors. A six-year technology evaluation and development program is foreseen, by the end of which an informed decision can be made on proceeding with demonstration of the ATW system.

Introduction

Commercial nuclear power stations in the United States will have generated nearly 87 000 tonnes (heavy metal) of spent fuel by 2015, even if there are no new plants built and no extensions of the operating licenses of existing plants. A system for fissioning the transuranics present in this fuel and transmuting certain selected long-lived fission products is being evaluated in a six-year study that began in October 1999. The design concepts for this system, known in the US as the accelerator transmutation of waste (ATW) system, are evolving. It is likely that the system will comprise an accelerator-driven subcritical assembly for accomplishing the desired transmutations while generating electric power, the revenues from which will serve to offset, in part, the costs of the system. An operating lifetime of 60 years has been chosen for initial systems studies.

An aqueous solvent extraction process has been selected as the baseline process for treatment of the LWR spent fuel. This selection was based on the need to process about 1 450 tonnes of spent fuel per year with high extraction efficiency for all of the actinide elements and with high decontamination of the uranium present in the spent fuel. The fuel to be processed will have cooled for at least 25 years from the date of reactor discharge. Owing to proliferation concerns commonly associated with conventional solvent extraction processes such as PUREX, a pyrochemical process is being evaluated as an alternative to PUREX. The alternative process must facilitate the recovery of 99.9% of the uranium and transuranic elements present in the fuel, and it must produce a separated uranium stream so that there is no carryover of fertile uranium into the transmutation cycle. In addition, it is required that the extracted uranium be of sufficient purity that it is disposable as low-level waste or qualified for surface storage in non-shielded containers.

The remaining fuel treatment processes in the ATW system are envisioned to be pyrochemical. The nitrate solution resulting from the first extraction cycle of the PUREX process is to be reduced to solid oxide and thence to metal. The metal product, containing all of the transuranic elements, some of the rare earth fission products, and all of the noble metal fission products, is then electrorefined to separate the transuranics. The TRU product is fabricated into metallic fuel subassemblies, which are inserted in the accelerator-driven transmuter array. These subassemblies are operated to burn-ups of the order of 33% and then discharged for processing to recycle the unburned transuranics and extract the newly generated long-lived fission products. A second pyrochemical process is utilised to perform these extractions. The key step in this second process is a chloride volatility separation. These pyrochemical processes will be described in detail in the following sections.

Factors driving process selection

The US accelerator transmutation of waste programme is motivated by the need to provide assurances that the high-level wastes from civilian nuclear power generation can be disposed of without significant risk to the public and to future generations. It also addresses the concern that nuclear fuel buried in a deep underground repository presents a potential source of weapons materials for future proliferators. It is constrained by the US policy regarding civilian use of plutonium; this policy presently avers that the US will not engage in reprocessing of spent fuel for the purpose of recovering plutonium for civil use. Excess weapons plutonium will be burned in selected reactors in the form of mixed oxide (MOX) fuel, but plutonium recycle is proscribed and the US nuclear utilities have been obliged to operate with the once-through fuel cycle. The current low price of uranium makes the once-through cycle economically favourable, but the prospects for timely emplacement of commercial spent nuclear fuel in a mined geologic repository are not particularly good.

An intensive study [1] of the performance of the proposed repository at the Yucca Mountain site in the state of Nevada concluded that the major contributors to the dose to the general population living 20 km from the repository thousands of years from now would be neptunium-237, iodine-129, and technetium-99, if the current policy of direct disposal is followed. This is attributed to the relatively high groundwater mobility of these species in the geologic structures in the vicinity of the repository. The source of ^{237}Np is the alpha-decay of ^{241}Am. Actinides other than neptunium were not significant factors in the offsite dose rate, due to their very limited solubility in water. However, a recent finding [2] that highly soluble Pu(VI) oxide may form readily in contact with water could have a remarkable effect on the predicted performance of the repository and make imperative the elimination of transuranics in general from material intended for repository disposal.

Approximately one year ago, the US Department of Energy initiated a study intended to set forth a course of action that could lead to implementation of an ATW system that would be capable of dealing with the inventory of commercial spent nuclear fuel accumulated by 2015, without consideration of spent fuel that might be produced by new plants constructed in the interim. The exercise was completed in late 1999 and duly reported [3] to the Congress of the United States, which had mandated the study. The baseline system for partitioning and transmutation recommended in this "roadmap" document is illustrated in Figure 1, where it is contrasted with the French version of the "double strata" fuel cycle [4]. The essential difference between these two approaches is the manner of dealing with the transuranic elements. In the double strata scheme, plutonium is multi-recycled, initially in light water reactors in the form of MOX fuel, and later in fast reactors. Because the PUREX process [5] separates pure uranium and pure plutonium but leaves the other transuranics, or minor actinides, in the stream with fission products, primarily the lanthanides, an additional aqueous separation process is required for the extraction of neptunium, americium and curium, which are directed to an accelerator-driven system for destruction. It may not be possible to complete this destruction without recycling the fuel fed to the ADS; a pyrochemical process has been postulated in Figure 1 for accomplishing this recycle step.

Figure 1. Comparison of French double strata system to the US ATW system

SNF: spent nuclear fuel; LWR: light water reactor; ADS: accelerator-driven system; ATW: accelerator transmutation of waste; MA: minor actinides (Np, Am, Cm,...); FP: fission products; Ln: lanthanides. Dashed lines represent material sent to high-level waste form production.

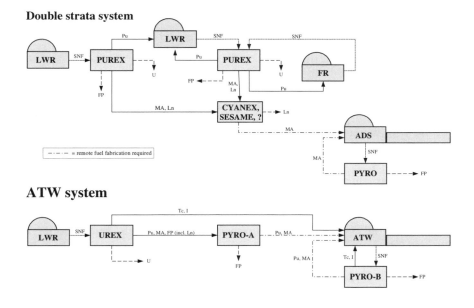

The ATW system is predicated on the policy position of the United States regarding civil plutonium utilisation. It is possible, however, within the constraints of this policy, to realise the energy potential residing in the transuranic elements present in the LWR spent fuel. This can be done by fissioning these elements in the fast neutron energy spectrum of an accelerator-driven subcritical assembly and generating electricity by utilisation of the thermal energy released. Therefore, all of the transuranic elements present in the LWR spent fuel are recovered and sent directly to the accelerator-driven system. A solvent extraction process was chosen for the head-end treatment of the LWR spent fuel, because it is desired to remove the bulk of the material present in the spent fuel (the uranium). Uranium removal is important for two reasons: it eliminates a fertile constituent that could breed more transuranics if fed to the transmuter, and it reduces the ultimate volume of the high-level waste produced by the overall process. The use of a solvent extraction process enables extraction of uranium at a purity level that would make it disposable as a Class C low-level waste or suitable for surface storage in an unshielded facility. Because the solvent extraction process envisioned is limited to uranium extraction only, it has been termed the "UREX" process. The head-end operations also include the extraction of iodine and technetium, which are intended for transmutation, probably to reside in the outermost regions of the subcritical assembly, similar to the radial blanket region in a fast reactor. A schematic flowsheet for the ATW treatment of LWR oxide spent fuel is shown in Figure 2.

Figure 2. Current reference flowsheet for the treatment of LWR oxide spent fuel for the accelerator transmutation of waste (ATW) system

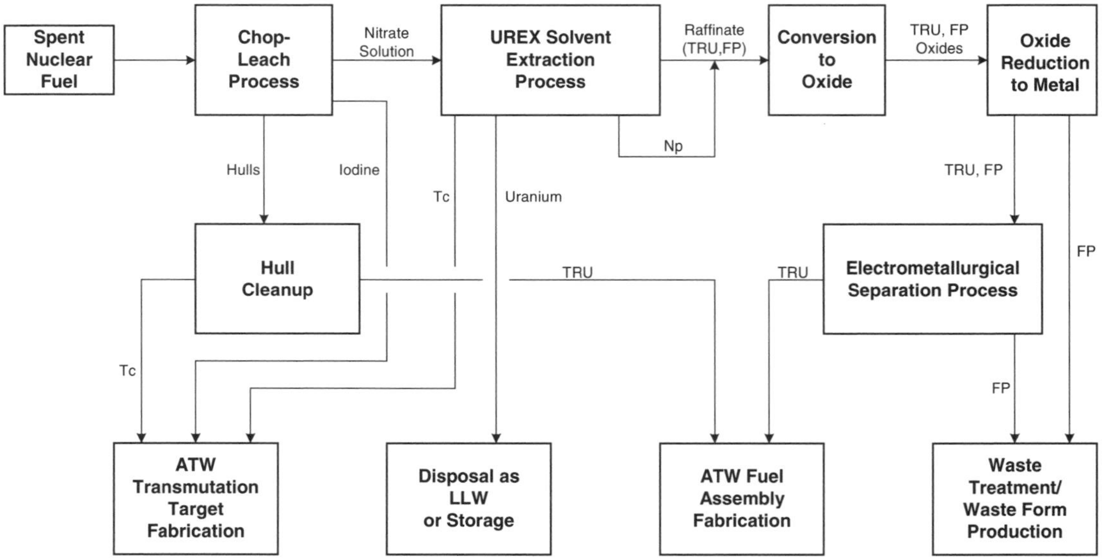

Removal of the uranium from the system greatly reduces the mass of material that must be carried through subsequent processing steps, from 1 450 tonnes heavy metal per year to 35 tonnes heavy metal per year. This makes the utilisation of pyrochemical processing very appealing, to capitalise on the compactness of the pyrochemical systems. Other features of pyrochemical processes are compelling as well, including the fact that the process proposed does not separate plutonium from the other transuranics. This tends to enhance the non-proliferation characteristics of the system, as part of a close-coupled system for the destruction of weapons-useable materials. Furthermore, the pyrochemical process recommended does not present the complication of a carryover of lanthanide fission products with the extracted transuranics, so no supplementary extraction processes are required. The very low volume of liquid wastes generated in the hybrid UREX/PYRO process was also a consideration in baseline process selection.

ATW system "PYRO-A" process

The head-end partitioning process presently envisioned for the ATW system calls for initial partitioning by means of a conventional aqueous solvent extraction process that is truncated at the first uranium extraction stage. That is, the uranium present in the LWR oxide spent fuel feed is removed quantitatively ($\geq$ 99.9% recovery) and the transuranic elements and fission products (other than xenon, krypton, tritium, iodine and technetium) are left in the raffinate solution. As shown in the schematic flowsheet in Figure 2, the raffinate is directed to an oxide conversion step where the nitrates are converted to solid oxides. The oxides are then reduced with metallic lithium and the metallic product is electrorefined to separate the transuranic elements from the fission products. The portion of the overall flowsheet that involves the conversion to oxide, reduction of the oxide to metal and electrometallurgical separation of the transuranics, has been termed the "PYRO-A" process.

A preliminary schematic process flowsheet for the PYRO-A process is shown in Figure 3. The UREX raffinate could be treated by a conventional oxalate precipitation/filtration/calcination step to produce the oxides of the transuranics and fission products, but this method would produce an additional waste stream that would have to be dealt with and may not facilitate the 99.9% overall recovery target set for the transuranic elements. Alternatives to this process are therefore being considered. Whatever the method chosen, the resultant oxide product would then be reacted with lithium metal in a lithium chloride carrier salt at a temperature of 650°C to reduce the oxides to the metallic state. Metallic lithium for use in the oxide reduction process is regenerated by electrowinning from the Li_2O-bearing LiCl carrier salt. The Group IA and IIA fission product elements, as well as some of the rare earth fission products, are not reduced and remain in the carrier salt; subsequently, these fission products are removed in the ceramic waste stream as described below. The remaining fission products and the transuranic elements are reduced to the metallic state and are transferred to an electrorefining vessel operating at 500°C with a LiCl-KCl electrolyte, where they are placed in the anode baskets. A more complete description of the reduction step is provided elsewhere [6].

Figure 3. Flowsheet for the ATW PYRO-A process for the treatment of UREX raffinate

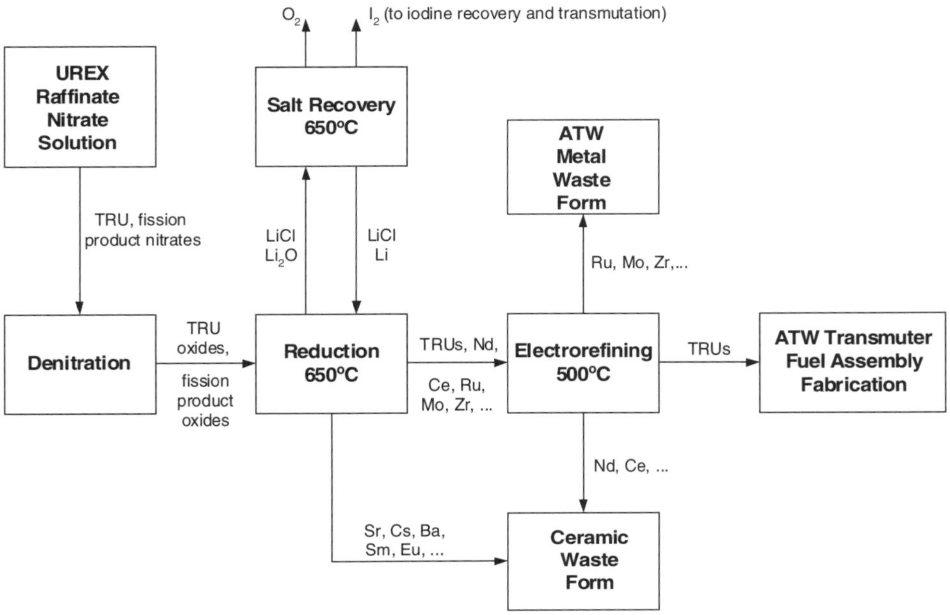

In the electrorefiner, the transuranic elements are electrotransported to a solid cathode. The elements that will form stable chlorides (e.g. Nd, Ce) will be dissolved in the electrolyte as chlorides, while those that are more noble (e.g. Ru, Mo, Zr) will remain behind in the anode basket as metals. Periodically, the electrolyte salt and the LiCl carrier salt are passed through a zeolite-packed column, where the fission products are extracted into the zeolite. The effluent salt is recycled, and the zeolite is mixed with glass frit and sintered to form the ceramic high-level waste form. The residue in the anode baskets is melted with suitable alloy additions to form the metal waste form (also a high-level waste). The cathode deposit of transuranic elements is removed from the electrorefiner, treated to remove the entrained electrolyte salt, and consolidated into ingots as feed for the fabrication of transmuter fuel assemblies.

ATW system "PYRO-B" process

The combination of metallic fuel with very high zirconium content and rather limited annual throughput requirements (~35 tonnes heavy metal per year) led to the selection of a pyrochemical process, referred to as "PYRO-B", for the treatment of the irradiated transmuter blanket fuel assemblies. The PYRO-B flowsheet is designed to extract the TRU elements (for recycle into fresh transmuter blanket fuel) and technetium and iodine fission products (for incorporation in transmutation assemblies) from spent transmuter blanket fuel and to provide waste streams that are compatible with either the ceramic (i.e. glass-bonded sodalite), or metallic (i.e. zirconium-iron alloy) waste form. The selection of pyrochemical processes for the treatment of the transmuter blanket fuel was based on their robust and compact nature, compatibility with the desired waste forms, and probable cost effectiveness. In contrast to the LWR fuel processing, high material throughput is not required for the treatment of spent transmuter blanket fuel, but the fuel will not likely have cooling times exceeding 6-12 months. Therefore, radiation levels and decay heat levels will be high; this does not pose a problem for pyrochemical processes, because the reagents used are stable and the equipment is designed for high-temperature application in the first place.

Two pyrochemical process options are being considered for treating irradiated transmuter blanket fuel: a chloride volatility process and an electrometallurgical process. The difference between the two options is the method by which the zirconium, the major component of the fuel, is removed from the transuranics and fission products. The baseline option for irradiated transmuter blanket fuel processing, as depicted in Figure 4, is based on a chloride volatility process (similar to the Kroll process) for zirconium extraction, coupled to an electrowinning process for TRU and fission product separation. Chloride volatility was chosen as the mechanism for TRU and zirconium separation because of the high zirconium content in the fuel and the existing industrial experience in zirconium metal production.

With the chloride volatility process, irradiated transmuter blanket fuel is removed from the transmutation system and allowed to cool for a reasonably short time; the fuel pins are separated from the fuel assembly hardware, and the pins are chopped. The chopped fuel is chlorinated, and the matrix zirconium from the fuel alloy, iron from the cladding, and the transition metal fission products (e.g. Tc, Ru, Mo) are vapour transported from the crucible containing the chlorides to a magnesium bath where the metal chlorides are reduced. The limited solubility of zirconium and the other transported metals in the magnesium allows for their separation from the magnesium/magnesium chloride mixture. The zirconium-based metal product is removed from the volatility processing system and any residual magnesium chloride left on the surface of the product is removed by vacuum distillation prior to sending the metal to the fuel fabrication process. Magnesium and chlorine are reclaimed by electrochemically decomposing the magnesium chloride produced during the reduction

Figure 4. Flowsheet for the PYRO-B process for the treatment of irradiated ATW transmuter blanket fuel

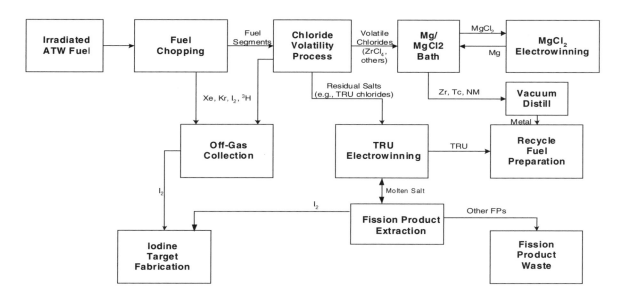

process. The remaining non-volatile metal chlorides (e.g. TRU, rare earths, Cs, Sr) are transferred from the volatility system to a molten salt (LiCl-KCl) bath in which the transuranics are electrowon from the solution and recycled to fuel fabrication. Periodically, the fission products are removed from the molten salt and converted to a stable ceramic waste form as described above. Iodine is removed from the molten salt, fabricated into targets and placed in transmutation assemblies. The technetium produced by the fission of transuranic elements in the transmuter blanket fuel remains in the zirconium stream and is recycled, without separation from the zirconium and the other noble metal fission products, into the transmuter system as part of the fresh blanket fuel.

The back-up electrometallurgical option involves the electrotransport of zirconium from an anode basket containing chopped ATW transmuter blanket fuel and deposition of the zirconium on a solid cathode. The transuranic elements partition to the molten salt electrolyte together with Group IA and IIA fission products and lanthanides. Technetium remains in the anode basket as a residue along with the other noble metal fission products. This anode heel is removed from the cell and incorporated in the fabrication of suitable transmutation target subassemblies. Iodine partitions to the molten salt as soluble metal iodides, which must somehow be collected and directed into the transmutation cycle. The transuranic elements are removed from the salt by electrowinning and deposited at a solid cathode that is periodically harvested. The extracted metallic TRU elements are cleaned of any adhering electrolyte salt and sent to transmuter fuel fabrication. The depleted electrolyte salt is then periodically treated to remove the rare earth and active metal fission products for immobilisation in the ceramic waste form.

ATW system waste form production

It was the intention during development of the conceptual baseline process flowsheet to avoid the generation of liquid wastes to the greatest extent possible. A significant volume of metallic waste, comprising the cladding hulls and other constituents of the LWR spent fuel assemblies, will be generated during the UREX operations preceding the solvent extraction steps. It is anticipated that these materials will be sent directly to the metal waste form preparation process. Subsidiary treatment of the cladding hulls will probably be required, for the purpose of maintaining the target actinide recovery level of 99.9%; a separate hull leaching treatment will create additional waste streams that must be accommodated.

While technically not a waste stream, the uranyl nitrate product of the UREX process will require further processing for its proper disposition. The solution will be treated by the same means developed for the corresponding solution of TRU elements and converted to a suitable oxide form (probably U_3O_8 if the option for disposal as a low-level waste is selected, and UO_3 if the material is to be stored for future use). Residual liquids from the uranium and TRU extraction steps are to be evaporated to dryness and any radioactive residue sent to the ceramic or metal waste streams.

The PYRO-A process waste streams lead to two high-level waste forms, one ceramic and one metallic, as seen in Figure 3. The fission products that do not tend to form stable chlorides (e.g. Ru, Re, Pd, Nb, Mo) will remain in the electrorefiner anode baskets, retained by the fine-mesh screens lining these baskets. The basket liners are held at temperature to allow salt to drain off, and are then combined with the cladding hulls (and, possibly, other fuel assembly constituents), melted, and cast into ingots. In the early stages of metal waste form development, a composition of zircaloy-8 wt.% stainless steel (Zr-8SS) was selected for use with zircaloy-clad fuel. The SS-15Zr composition selected for use with stainless steel clad fuel is now a well-characterised waste form material with well-understood properties [7]; it is being prepared in the course of the demonstration of the electrometallurgical treatment process demonstration with sodium-bonded fuel and blanket assemblies discharged from the EBR-II reactor in Idaho. Development of the Zr-8SS waste form has not proceeded beyond the initial studies, which showed Zr-8SS properties and behaviour are quite favourable for application as a waste form for the ATW system. However, development work is still needed to bring Zr-8SS technology to an appropriate level of maturity. The microstructure of this alloy composition, shown in Figure 5, appears to be reasonably stable. With the large excess of zirconium, the comparatively small concentration of noble metal fission products tends to be localised in the primary α-zirconium phase and not at phase boundaries. Limited corrosion studies with this composition, using representative concentrations of (non-radioactive) fission product elements, suggest that the alloy will possess excellent corrosion resistance in a typical repository environment.

Figure 5. Backscattered electron image of Zr-8SS alloy. The noble metal fission products tend to reside in the primary α-zirconium phase.

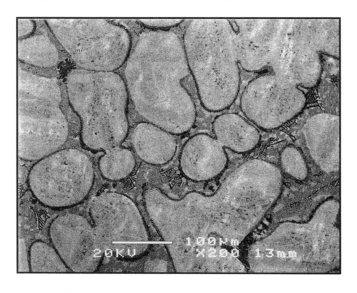

Again referring to Figure 3, salt-borne fission products in the PYRO-A process derive from the oxide reduction step and from the electrorefining process. The salt stream from the oxide reduction step must be combined with sufficient KCl to make the salt compatible with the electrorefiner salt stream and the subsequent processing steps. The two salt streams are then combined and passed through a zeolite column, where the fission products in the salt are extracted into the zeolite structure. The salt-loaded zeolite is removed from the column when the limiting heat loading (due to decay heating by the sorbed radionuclides) is reached. The salt-loaded zeolite is crushed, mixed with glass frit, and the mixture heated above the softening point of the glass (i.e. to 850°C) to produce a solid monolith. During this heating process, the zeolite transforms to sodalite. Much of the salt and its solutes remain within the sodalite lattice, but the rare earth fission products and any trace actinide elements will tend to form stable secondary phases, generally oxides or oxychlorides. The resultant glass-sodalite composite ceramic waste form has been shown [8] to have good corrosion and leach resistance. There is more experience with the environmental behaviour of this waste form than with the metal waste form, so recent development work has focused on reducing waste volume and minimising fabrication costs. Figure 6 shows a large section of the ceramic waste form; such sections have been prepared by ambient-pressure sintering of zeolite containing non-radioactive fission product elements at concentrations up to 5 wt.% and with 25% glass binder. It appears that the fraction of glass can be reduced further.

Figure 6. Sample of simulated ceramic waste form prepared by sintering at ambient pressure at 850°C. Inset shows microstructure of glass-sodalite composite.

The pyrochemical processes that make up the PYRO-B flowsheet (Figure 4) are characterised by low amounts of secondary waste generation. This is a key advantage of this family of process options. Figure 4 does not show the detail of waste form production, but the process involves the same two waste streams and method of production encountered with the PYRO-A process. If an electrorefining alternative were to be used instead of the chloride volatility step, it is possible that a mixed chloride/fluoride electrolyte might be required. This would change substantially the production of the ceramic waste form, because the zeolite system for immobilisation of fission products is not compatible with fluoride ions. Prior research on the processing of other fuel types (e.g. molten fluoride salt fuels) led to the development of a fluorapatite waste form with excellent properties; it is not certain, however, that the waste volume minimisation objectives could be met with this process variation.

Recycle of the zirconium extracted in the course of transmuter blanket fuel processing could significantly reduce the volume of the metal waste form produced. Using conservative estimates for fission product loading of the waste forms, the processing of 1 450 tonnes (heavy metal content) of LWR spent oxide fuel per year for transmutation would produce nearly 400 cubic meters of high-level waste annually, assuming no recycle of zirconium. This amounts to about 0.28 cubic meters per tonne of fuel processed, which is comparable to the best results achieved to date at contemporary PUREX reprocessing plants. It is quite likely that this value could be reduced to 0.1 m^3/t after a modest experimental effort at process optimisation. For purposes of comparison, the unpackaged volume of 1 450 tonnes of LWR spent fuel is about 600 cubic meters, or approximately 0.4 m^3/t. High-level waste volume is considered to be a significant metric, because a volumetric reduction in waste volume is directly reflected in avoidance of the cost of fabricating, loading, handling and emplacing repository disposal containers.

Acknowledgement

This work was supported by the US Department of Energy under Contract No. W-31-109-ENG-38. It was performed at the Argonne National Laboratory, operated for the US Department of Energy by the University of Chicago.

REFERENCES

[1]   United States Department of Energy, Office of Civilian Radioactive Waste Management, "Viability Assessment of a Repository at Yucca Mountain", DOE/RW-0508, December 1998.

[2]   J.M. Haschke, T.H. Allen and L.A. Morales, "Interaction of Plutonium Dioxide with Water: Formation and Properties of PuO_{2+x}", *Science* 287, pp. 285-287, 14 January 2000.

[3]   United States Department of Energy, "A Roadmap for Developing Accelerator Transmutation of Waste (ATW) Technology, A Report to Congress", DOE/RW-0519, October 1999.

[4]   OECD Nuclear Energy Agency, "Actinide and Fission Product Partitioning and Transmutation: Status and Assessment Report", OECD, Paris, 1999.

[5]   M. Benedict, T.H. Pigford and H.W. Levi, "Nuclear Chemical Engineering", McGraw-Hill, New York, 1981.

[6]   E.J. Karell, K.V. Gourishankar, L.S. Chow and R.E. Everhart, "Electrometallurgical Treatment of Oxide Spent Fuels", International Conference on Future Nuclear Systems, Global'99, Jackson Hole, Wyoming, 29 August-3 September 1999.

[7]   D.P. Abraham, Metal Waste Form Handbook, Argonne National Lab. Report ANL-NT-121, June 1999.

[8]   J.P. Ackerman, C. Pereira, S.M. McDeavitt and L.J. Simpson, "Waste Form Development and Characterization in Pyrometallurgical Treatment of Spent Fuel", Proceedings, Third Topical Meeting on DOE Spent Nuclear Fuel and Fissile Materials Management, p. 699, American Nuclear Society, Charleston SC, 8-11 September 1998.

OVERVIEW OF RIAR ACTIVITY ON PYROPROCESS DEVELOPMENT AND APPLICATION TO OXIDE FUEL AND PLANS IN THE COMING DECADE

A.V. Bychkov, O.V. Skiba, S.K. Vavilov, M.V. Kormilitzyn, A.G. Osipenco
State Scientific Centre of Russian Federation
Research Institute of Atomic Reactors
Russia, Dimitrovgrad, 433510
Fax: +7 (84235) 35648, E-mail: bav@niiar.ru

Abstract

A brief description of the activities of the Research Institute of Atomic Reactors (RIAR) in the field of pyroelectrochemical method development is given. A description of basic processes for production and reprocessing of oxide fuel in molten salts is adduced. A brief overview of the basic direction of activities and schedules over the next 10 years with regard to development of pyroelectrochemical technology is submitted.

Introduction

Global awareness of problems associated with nuclear energy has been a determining factor in the application of safe and highly efficient technologies, with the result that the following principles have become increasingly important:

- The *closed cycle*, i.e. internal isolation of technological processes, aimed at reducing a gross output of dangerous substances which are harmful to the environment.

- *Optimisation* of technological systems which is intended for achieving necessary results (technological and commercial) with the maximal exception of excessive stages.

- Maximum level of internally *inherent safety*, i.e. using processes in which safety is based on its own "natural" properties of technological system, creating a low degree of ecological damage probability.

The nuclear fuel cycle (NFC), as a complex technological system, should be developed on the same principles. At a 1997 IAEA symposium the following thesis was established. The fuel cycle facilities of tomorrow are driven by the need to achieve:

- Demonstrably high levels of safety.

- Minimum fuel cycle costs.

- Minimum environmental impact, including minimum waste generation.

- Maximum utilisation of natural resources.

- Minimum proliferation risk and maximum safeguards visibility.

- Diversity and security of energy supply.

Non-aqueous methods are possible new technologies for NFC. An integrated flowsheet of known pyroelectrochemical method is given in Figure 1. Both flowsheets meet the above requirement.

Advanced fuel cycles developed in RIAR (Dimitrovgrad) are based on the following elements:

- Pyroelectrochemical processing of the spent oxide fuel for production of the fuel in granulated form.

- Use of the fuel rods with vibro-compacting fuel.

- Organisation of the fuel recycle stages as a remotely controlled production line.

Advantages of the pyroelectrochemical fuel technology included in this fuel cycle are as follows:

- High chemical and radiation resistance of the solvents.

- No neutron moderators.

- High content of fissile materials in electrolyte (> 30% wt.%).

- Non-proliferation resistance (high gamma activity of reprocessed fuel).

- Process cycle can be carried out in one apparatus.

- Minimised volume of high level wastes.

- Final product, oxide fuel granules, are used directly for vibro-compacting into a fuel pin.

Basis of pyrochemical reprocessing of oxide fuel

Pyrochemical technology is based on the production of actinide oxides from molten chloride systems. Physical and chemical laws of the processes occurring in the molten alkali chlorides with U and Pu have been re-examined in more detail. Information on the behaviour of the main fission products and impurities in molten chlorides is also available. Some data on formal electrode potentials of elements in molten chlorides are given in Table 1.

Table 1. Formal electrode potentials of actinides and fission products in molten chlorides

Me(n)/Me(m)	$E^{*}_{Me(n)/Me(m)}$, V (ref. chlorine electrode)		
	3LiCl-2KCl $T = 773$ K	NaCl-KCl $T = 1\ 000$ K	NaCl-2CsCl $T = 873$ K
Sm(II)/Sm	-3.58	-3.44	-3.58
Eu(II)/Eu	-	–	-3.39
Pr(III)/Pr	-3.11 (723 K)	–	–
La(III)/La	-3.07	–	–
Am(II)/Am	-3.05	–	–
Ce(III)/Ce	-3.04	–	-3.08
Nd(III)/Nd	-3.02	–	–
Pu(III)/Pu	-2.82	-2.64	-2.83
Np(III)/Np	-2.315	–	–
U(III)/U	-2.31	-2.19	-2.39
Zr(IV)/Zr	-2.08	-1.98	-2.17
Cr(II)/Cr	-1.65 (723 K)	-1.59	–
Np(IV)/Np(III)	-1.51	–	–
U(IV)/U(III)	-1.40	-1.39	-1.45
Fe(II)/Fe	-1.38	-1.34	-1.48
Nb(III)/Nb	-1.35	-1.13	–
Ni(II)/Ni	-1.01	-0.97	–
Ag(I)/Ag	-1.06 (723 K)	-0.838	-0.932
Mo(III)/Mo	-0.90	-0.823	-0.97
UO$_2$(VI)/UO$_2$	-0.50	-0.40	-0.65
Pd(II)/Pd	-0.536 (723 K)	-0.37	-0.48
Rh(III)/Rh	-0.51	–	-0.44
Ru(III)/Ru	-0.43 (723 K)	–	-0.413 (783 K)
NpO$_2$(V)/NpO$_2$	-0.32 (723 K)	–	–
Pu(IV)/Pu(III)	+0.01	+0.09	-0.05
PuO$_2$(VI)/PuO$_2$	+0.28	+0.34	+0.12

Concerning on oxide fuel reprocessing technology the following properties of U and Pu are more important:

- From an electrochemical point of view, U and Pu oxides behave like metals. During dissolution in the molten salt or anode oxidation they are forming the complex oxygen ions MeO_2^{n+}, which are reduced on cathode up to oxides.

- Under high temperatures (> 400°C) UO_2 and PuO_2 have electric conduction. Thus, while electrolysis the formation of crystals and increase of cathode deposit are possible. The lower electric conduction of oxides in comparison with the melt provides the stable flat front of crystallisation and allows obtaining compact cathode deposits.

- In the molten alkali chlorides uranium has the stable ions U(III), U(IV), U(VI). At the same time the highest states of Pu oxidation Pu(V) and Pu(VI) are stable only in the definite field of ratios for oxidation reduction potentials of the system. So during the joint electrodeposition of UO_2 and PuO_2 it is necessary to treat the melt with a chlorine-oxygen gaseous mixture for creation of the ion PuO_2^{2+} required concentration.

- From any oxidation state plutonium can be conversed into oxide by changing of the oxidation-reduction potential of the system. This process is called "precipitative crystallisation". Under oxidation conditions uranium stays in the melt and thus it is possible to fractionally separate the plutonium from nuclear fuel.

- UO_2 and PuO_2 are reduced in electropositive area. The majority of FPs is reduced at more negative potentials, so in the process of UO_2 and PuO_2 electrodeposition their purification from impurities occurs.

The data on chemistry of U, Pu and FPs were the basis for development of technologies for production of fuel compositions and reprocessing of spent nuclear fuel.

Main pyroelectrochemical processes

Technological processes for production of oxide granulated fuel from the molten alkali chlorides have been developed. The initial products of the process are Pu and U oxides in the form of powder. However, it is being considered to use different types (of isotopic and chemical composition) of initial materials.

Electrodeposition of uranium dioxide

The process is carried out using molten $NaCl-2CsCl$ (or $NaCl-KCl$)-UO_2Cl_2, prepared by chlorination of uranium oxides. During electrolysis the formation of UO_2 dense cathodic deposits occurs. The special electrolysis modes allow to obtain a product with ra ecovery rate of 99.0-99.5% for manufacturing of vibro-compacted fuel pins with high effective density (more than 9.0 g/cm^3). The technology of crystal uranium fuel production has passed the complete cycle of development from laboratory installations up to creation of semi-industrial facility.

In the beginning of the 70s, experiments with irradiated UO_2 were carried out at the BOR-60 reactor. The experiments demonstrated that pyroelectrochemical methods allow to reprocess spent nuclear fuel (burn-up of 7.7% h.a., cooling period of six months) with a high recovery rate (99%) and satisfactory decontamination from FPs (DF = 500-1 000). After experiments, reprocessed UO_2 was irradiated repeatedly in the BOR-60 reactor.

Precipitative crystallisation of plutonium dioxide

The process is carried out using molten NaCl-KCl after chlorination/dissolution of PuO_2 or metal Pu. During treatment of the melt with a gas mixture containing chlorine and oxygen, the formation of crystalline PuO_2 occurs. The process of production of granulated PuO_2 is developed and is tested in three options:

- Production of crystalline PuO_2 from reprocessed "power" grade plutonium with purification from americium.

- Extraction of crystalline PuO_2 from the irradiated MOX fuel. The process is tested on the spent MOX fuel of BN-350 and BOR-60 reactors.

- Conversion of military origin Pu alloy into dioxide.

Electro-codeposition of uranium and plutonium oxides

This process includes the operation of initial oxides chlorination in molten NaCl-2CsCl with formation of uranylchloride and plutonium chlorides and the operation of oxidation for transformation of Pu into plutonylchlorides. Electrolysis is carried out with treatment of melt by chlorine-oxygen gas mixture, causing UO_2 and PuO_2 codeposition to occur. The process is based on the discharge of UO_2^{2+}, PuO_2^+ and PuO_2^{2+} ions on the cathode from molten NaCl-2CsCl; under electrolysis the UO_2 deposition rate is pre-set by the current density; the PuO_2 deposition rate is limited by the diffusion of PuO_2^+ and PuO_2^{2+} to the cathode surface. The content of plutonyl ions in the melt is regulated by oxidation of the melt by gas Cl_2-O_2 mixture. The cathodic deposits obtained are quasi-homogeneous crystal system $(U,Pu)O_2$. It has a two-phase composition: solid solution of PuO_2 in UO_2 crystals and solid solution of UO_2 in PuO_2 crystals. It usually ranges from 80-87% from the theoretical density and depends on the electrolysis condition and U to Pu ratio.

After removal of captured salts and crushing, the product is used for vibro-compacted fuel pins manufacturing. The fuel recovery rate (with recycled products) is 98.5-99.5%. The technology was developed and the semi-industrial facility for production of MOX fuel was constructed. This facility has been in operation since 1988. It provides manufacture of fuel for the BOR-60 reactor and experimental FAs for the BN-600 reactor.

The obtained products have a particle density close to the theoretical (10.7-10.9 g/cm^3), which allows to supply high effective density of the fuel column during vibro-compacting.

The necessary equipment for realising the technological processes in remote conditions was developed. The head device – chlorator-electrolyser – uses pyrolitic graphite as the material for the bath-anode, cathode and gas tubes. Some types of chlorator-electrolysers are tested for fuel processing which differ only in crucible volume. The device currently used has a working volume of 40 l and is designed for loading 30 kg of initial material, that is multiple by one FAs for BN-600 reactor. The device is shown in Figure 2. Apart from the chlorator-electrolyser, other equipment for crushing of cathodic deposits, for removal of captured salts from fuel, for drying and classification are used.

Military plutonium conversion to MOX fuel

For the last few years, the RIAR has been working toward the adoption of the MOX fuel production flowsheet for conversion of military plutonium into MOX fuel for fast reactors. This flowsheet has some differences from the one previously discussed. A military plutonium-gallium alloy is used as the initial material, so the process is adopted to other chlorination mode. The demonstration programme, which consisted of converting 50 kg real military plutonium to MOX fuel, was undertaken from 1998 to 1999. The modified flowsheet allows to obtain MOX fuel from plutonium of military origin for the BOR-60 and BN-600 fast reactors.

The process described above provides the programme of fuel pin manufacture with vibro-compacted oxide uranium and MOX fuel for operation of the BOR-60 reactor and for tests in the BN-600 and BN-350 fast reactors (Table 2).

Table 2. Oxide fuel production and reprocessing by pyrochemical methods in RIAR

Type of fuel	Amount, kg	Production period, year	Reactor
Pyrochemical production			
UO_2	1 696	1976-2000	BOR-60
	739	1983, 1993	BN-350
	120	1988	BN-600
	535	1988-1989	RBT-10
PuO_2	100	1980-1982	BOR-60
$UPuO_2$	986	1983-2000	BOR-60
	75	1984	BN-350
	465	1987, 1990, 2000	BN-600
	277	1989-1992	BFS

Type of fuel	Burn-up	Weight, kg	Year	Reactor
Pyroelectrochemical reprocessing tests				
UO_2	1%	3.3	1968	VK-50
UO_2	7.7%	2.5	1972-73	BOR-60
$UPuO_2$	4.7%	4.1	1991	BN-350
$UPuO_2$	21-24%	3.5	1995	BOR-60

New fuel composition production from molten salt

The application of the unique properties of UO_2 under electrolysis of molten salt other fuel compositions were produced by pyroelectrochemical method. These are the following:

- MOX fuel with high plutonium content (30-50% of Pu).

- LWR type MOX fuel containing 2-8% of Pu.

- Oxide (UO_2 and MOX) fuel containing 3-9% of Np.

- Oxide (UO_2 and MOX) fuel with americium additions (1-5% of Am).

- Mixed uranium-thorium oxide fuel.

In the framework of the programme for conversion of military plutonium the PuO_2 powder was produced for the future production of MOX fuel pellets.

Pyrochemical reprocessing of irradiated oxide fuel

The following options were considered and are now under development for reprocessing of fast reactor irradiated fuel at RIAR:

- Reprocessing of uranium fuel with production of UO_2 for recycle.

- Reprocessing of MOX fuel with plutonium recycle as the most valuable component.

- Reprocessing of MOX fuel with production of MOX fuel.

Pyroelectrochemical process of PuO_2 extraction from core spent nuclear fuel was developed for demonstration of closed fuel cycle of the BOR-60 reactor.

The thermal or mechanical decladding of fuel is considered a first stage. The operation sequence of pyroelectrochemical reprocessing in chlorator-electrolyser is submitted in Figure 3.

- Fuel chlorination in molten NaCl-KCl or LiCl-NaCl-KCl-CsCl. The complete dissolution of fuel components is taking place.

- Electrolysis for removal of part of UO_2 free from Pu. There some FPs (Zr, Nb, Ru, Rh, Pd, Ag) are captured by cathodic deposit. U/Pu separation factor at this stage is no more than 120-140.

- Precipitative crystallisation of PuO_2 decontaminated from FPs, allowing to obtain crystalline PuO_2 ready for vibro-compacting. Most of the Pu (99.5-99.9%) is collected in bottom PuO_2.

- Additional electrolysis for removal of uranium. The main number of FPs are deposited simultaneously with UO_2 from melt.

- Molten salt purification from impurity is carried out by introduction of sodium phosphate into the melt. As a result the formation and precipitation of impurity phosphates occur, which are insoluble neither in molten chlorides nor in water. Only Cs, Rb and, partially, Sr remain in molten salt.

Two integrated tests were completed during the 90s: (1) with the BN-350 MOX-fuel and (2) with the BOR-60 high burn-up fuel. During the second test two FAs of the BOR-60 reactor with burn-ups of 21% and 24% h.a were reprocessed. It was carried out in 1995 and gave new data for future development of pyroprocessing. The decontamination of Pu is quite sufficient from the point of view of fuel nuclear-physical properties of fuel (Table 3).

Table 3. PuO_2 decontamination factors (DF) from main FPs

Test /FP	Ru-Rh	Ce-Pr	Cs	Eu	Sb
DF for BN-350 test	50	220	> 3 000	40	200
DF for BOR-60 test	33	40 ÷ 50	4 000	40 ÷ 50	120

The study on radiation conditions during experiment show that any appreciable "disturbance" in normal work of RIAR had not taken place.

DOVITA programme

Taking into account advantages of pyrochemical technology and vibro-compacting method, RIAR began research under the DOVITA programme (Dry reprocessing, Oxide fuel, Vibro-compact, Integral, Transmutation of Actinides) in 1992. It is the programme for demonstration of minor actinide burner reactor (Figure 4). The reprocessing of the irradiated fuel and targets is offered in two ways: a) pyroelectrochemical and b) non-complete vacuum-thermal. The vibropac technology is applied to the manufacture of fuel pins. All products are reprocessed with the purpose of complete recycle of Pu, Np, Am and Cm. A number of experimental researches on Np and Am recycle were fulfilled. The fuel with the Np additives is already irradiated in the BOR-60 reactor up to burn-up 13% h.a., and the fuel with the Am additives is prepared for irradiation.

Current activities and main future programmes

RIAR continues to work in the field of the pyroprocess development. The main goal of these various experiments is to introduce this technology into nuclear industry. Over the next decade (2000-2010) the following programmes related to pyroelectrochemical process will be covered:

- Demonstration integrated experiments on pyroelectrochemical reprocessing of irradiated UO_2 and MOX FAs of BOR-60 reactor under the flowsheets $UO_2 \rightarrow UO_2$ and MOX$\rightarrow$MOX. They are planned for 2000-2001. The aims of these tests are the demonstration of the BOR-60 closed fuel cycle and the development of the flowsheet for next level demonstration work.

- Programme on creation of the BOR-60 reactor closed fuel cycle on basis of pyroprocessing and vibro-compacting. This programme began in 1999, and included the complex of modernisation works for the pyrochemical facility. The main task is to prepare conditions for works with the BOR-60 irradiated fuel. The proposed date for closing of the BOR-60 fuel cycle is 2003.

- Development and semi-industrial demonstration of military origin plutonium conversion process into MOX fuel of FBR: BOR-60 and BN-600. This programme was started in 1998 and will continue up to 2005-2007.

- Conceptual and experimental investigations of the external fuel cycle for the actinide burner fast reactor using the pyrochemical processes (the DOVITA programme). This scientific programme will be continued up to experimental demonstration of all its stages.

- Investigations and demonstration tests on separation of minor actinides from fission products in pyrochemical technology.

- Development of the highly efficient equipment for the pyroelectrochemical reprocessing of the LWR MOX fuel

- Investigations and demonstration tests on an immobilisation of pyrochemical wastes.

The realisation of the above-mentioned programmes would provide data for industrialisation of pyroelectrochemical method in the future.

Figure 1. Closed fuel cycles based on pyrochemical processes developed by ANL (right branch) and RIAR (left branch)

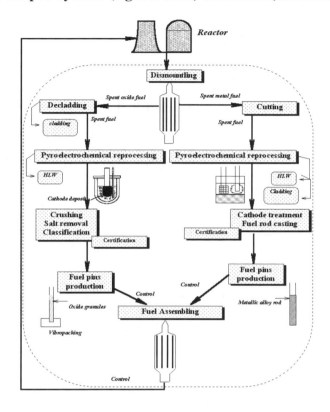

Figure 2. Chlorator-electrolyser (stages of fuel chlorination and precipitative crystallisation)

1 – cover, 2 – flange, 4 – mixing assembly, 5 – working vessel,
6 – defence vessel, 7 – furnace, 8 – pyrographite bath, 9 – gas supply tube

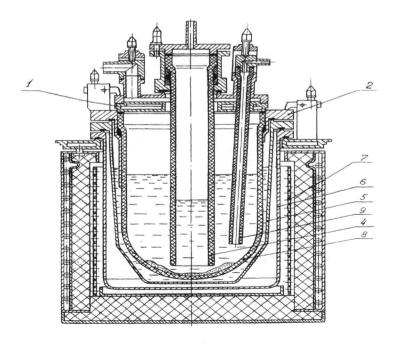

Figure 3. Flowsheet of pyroelectrochemical reprocessing (test with the BOR-60 fuel)

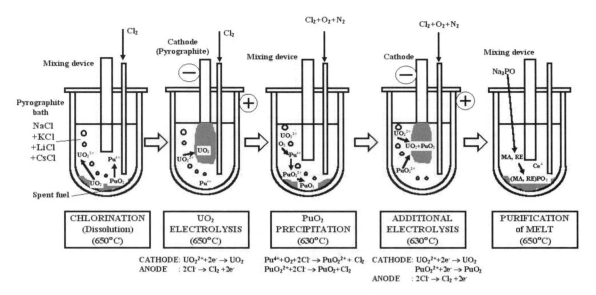

CHLORINATION (Dissolution) (650°C)	UO₂ ELECTROLYSIS (650°C)	PuO₂ PRECIPITATION (630°C)	ADDITIONAL ELECTROLYSIS (630°C)	PURIFICATION of MELT (650°C)

CATHODE: $UO_2^{2+}+2e^- \rightarrow UO_2$
ANODE : $2Cl^- \rightarrow Cl_2 +2e^-$

$Pu^{4+}+O_2+2Cl^- \rightarrow PuO_2^{2+} + Cl_2$
$PuO_2^{2+}+2Cl^- \rightarrow PuO_2+Cl_2$

CATHODE: $UO_2^{2+}+2e^- \rightarrow UO_2$
$PuO_2^{2+}+2e^- \rightarrow PuO_2$
ANODE : $2Cl^- \rightarrow Cl_2 +2e^-$

Figure 4. Proposed fuel cycle for reactor-burner of minor actinides (DOVITA)

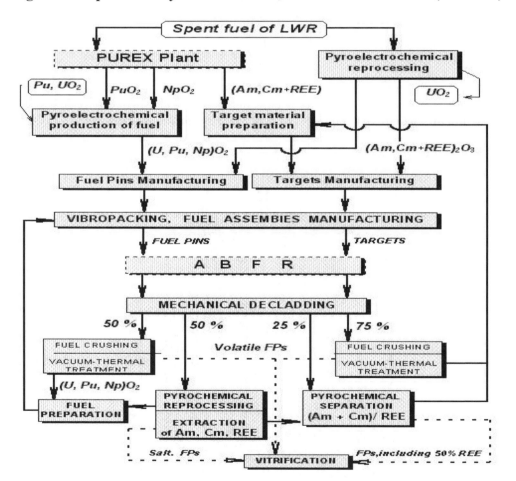

BNFL'S MOLTEN SALTS PROGRAMME: INDUSTRIALISING PROCESSES FOR IRRADIATED FUEL

**Rob Thied*, David Hebditch, Peter Parkes,
Richard Taylor, Bruce Hanson, Peter Wilson**
*British Nuclear Fuels, Sellafield, Seascale, Cumbria, CA20 1PG
E-mail: rct2@bnfl.com

Abstract

Molten salt processes for irradiated nuclear fuel have been studied internationally for several decades, notably at Argonne National Laboratory in the US and at the Research Institute for Atomic Reactors, Dimitrovgrad, Russia, where pilot scale demonstration facilities are operational for fast reactor fuels. There are ongoing major studies of partitioning and transmutation of thermal reactor generated wastes, for example by JNC/CRIEPI (OMEGA), CEA/COGEMA (SPIN), USDOE (ATW), JRC and OECD/NEA, that feature or are turning towards pyrochemical processing. The ongoing studies in Japan and the US in particular are broadening the remit of partitioning and transmutation to integrate these with advanced fuel cycle including reactor or accelerator with a fast neutron flux. As part of the European Commission 5th Framework Programme on partitioning and transmutation, pyrochemical processes are proposed to receive considerable research effort within European laboratories. Potential overall applications of pyrochemical processing include conditioning of irradiated fuel, low DF recycle of high burn-up thermal MOX and fast reactor fuels, and partitioning for transmutation.

BNFL believes that one of the greatest challenges facing the technology will be to convert it into commercially viable industrial-scale processing. BNFL has a well proven track record of turning science into technology, and its current programme on molten salts is geared towards identifying, understanding and resolving the uncertainties in existing processes with a view to industrialisation and commercial application. The project is aimed in breadth at a complete scheme of fuel treatment, from the head end through a separation process to the immobilisation of salt wastes, and in depth from basic chemistry and materials science to engineering design studies. In accordance with BNFL's innovation policy, the overall project is business funded with assessment of potential value and risk at each stage of development.

The present paper gives a general view of BNFL's molten salts project over the last decade and as intended for the future, in particular how it contributes towards the EC 5th Framework Programme investigating partitioning and transmutation.

Introduction

Processing of nuclear fuel has, by long-established tradition, been performed by solvent extraction, which has the following advantages over previous precipitation methods:

- Lends itself easily to large-scale multi-stage operation.

- Provides the high decontamination factors required for subsequent manual handling of the products under industrial conditions.

- Involves few mechanical handling and solid-liquid separations.

- Is suitable for continuous, remote operation.

After early experience with various solvents, the commercial industry has plumped without exception for the PUREX process in one or other of the many possible variations on the basic theme. Here fuel is dissolved in nitric acid, extracted with tri-*n*-butyl phosphate (TBP) in an inert diluent to separate uranium and plutonium from fission products and other impurities, and backwashed successively with an aqueous reducing agent and very dilute nitric acid to separate plutonium from uranium. Despite some difficulties with ancillary processes, operating experience on the main line is generally excellent, with decontamination factors of 10^6 to 10^8 achieved as a matter of routine.

However, there are certain drawbacks, increasingly recognised in recent years.

- In starting with a solid input and eventually delivering solid products from an essentially liquid-phase operation, the process requires dissolution and solidification stages that might ideally be avoided, and generates considerable quantities of aqueous and solvent waste that themselves have to be decontaminated or converted into a solid form with a view to disposal.

- The high decontamination factors are unnecessary for products that are intended for recycle into reactors where remaining impurities would almost immediately be outweighed by those newly generated.

- The virtual elimination of fission product radioactivity would remove an element of physical danger and risk of discovery from any attempt to divert the products from civil to military or terrorist use, and so requires a high degree of supervision and physical security in storage facilities.

- The aqueous waste does not lend itself easily to separation of long-lived radioactive components if this should be required for separate disposal or transmutation to shorter-lived and stable species.

Accordingly dry (pyrochemical) processes have lately received much attention in the hope of improving especially economy, waste management and resistance to weapon proliferation.

Status of international pyrochemical work

Several of the very early nuclear reactors were fast flux, breeder type such as the Experimental Breeder Reactor I (EBR-I) in the US that produced power and proved breeding of fissile material in the early 1950s. Pilot-scale reprocessing of fast reactor fuels, using both PUREX or pyrochemical technologies, demonstrated closure of the fuel cycle. Benefits of fast flux include high fuel burn-up, reduced sensitivity to plutonium isotopic quality, breeding gain, tolerance to low decontamination factor for recycled fuel and fission of recycled actinides. R&D continues in pyrochemical processing of irradiated fuel and conditioning of thermal and fast fuels for disposal. Emphasis has turned to partitioning of HLW and irradiated fuels for waste management purposes.

Early reactors used metallic fuels but these swelled excessively at low burn-ups. Oxide fuel began to predominate. In the 1970s, Argonne National Laboratory (ANL) in the US found that 75% smear density cast metallic fuel rods with a sodium bond could achieve high burn-ups. Early ANL fuels from EBR-II were treated by a simple pyrometallurgical route, mainly volatilisation and slagging, that was unsuitable for larger programmes. This preceded the integral fast reactor concept using metallic fuel alloyed with zirconium. Fuel is recycled after short cooling by electrorefining metallic uranium in molten salt and remote recasting of "hot" metallic fuel rods. The process inherently collects minor actinides with plutonium that is never completely separated from uranium, and rejects lanthanides to a degree adequate for fast-flux operation. The non-proliferation characteristics are favourable and actinide species are recycled. Subsequently ANL developed electrometallurgical treatment as a waste conditioning method and has just treated > 1 t (HM) of EBR-II fuel.

The former Soviet Union and now Russian programme at RIAR Dimitrovgrad has successfully run a small group of commercial generating oxide-fuelled FBRs and has demonstrated recycling of fuel at tonnage scale between fast reactors using a pyrometallurgical process. Like the ANL system, this aimed for technical simplicity and low DFs whilst facilitating economical remote fabrication of "dirty" fuel with vibropac technology. The technology uses dissolution of mixed oxide in molten alkali salt and electrorefining and precipitation of heavy metal oxides. Apart from lanthanides, the DFs of recycled fuel from the ANL and RIAR processes tend to be around 10^3.

Several major international studies of partitioning and transmutation of thermal reactor generated wastes are underway, for example by JNC/CRIEPI (OMEGA), CEA/COGEMA (SPIN), USDOE (ATW), JRC and OECD/NEA. Earlier studies concluded that owing to the projected good safety of planned geological disposal and the costs and doses associated with additional processing, partitioning and transmutation was not justifiable. The ongoing studies in Japan and the US in particular are tending to broaden the remit of partitioning and transmutation to integrate these with advanced fuel cycles including reactors or accelerator-driven systems with a fast neutron flux.

Partitioning schemes aim to separate specific hazardous and long-lived radioisotopes, such as minor actinides and ^{129}I and ^{99}Tc, from fuel solutions or high level wastes, and specifically from nuclides such as the higher lanthanides that would interfere with subsequent transmutation. Some add-on techniques that separate minor actinides, are compatible with existing PUREX technology. Alternatively pyrochemical processes may be used to treat PUREX reprocessing wastes, as for example in the CRIEPI double stratum concept. The advantage of adding to PUREX processes is that of using existing technology and treating existing wastes. The disadvantages of the incremental approach are (a) economic, that is adding to PUREX costs, and (b) not addressing advanced fuel cycle and reactor types. There is an international consensus developing that such pyrochemical processes should be researched more intensively to produce an advanced fuel cycle with a synergistic approach to the selection of fast-flux reactor, fuel and process type. Goals include providing minimal adequate separations, partitioning of key species to fuel or targets as an inherent part of the technology, simplified fuel fabrication and avoiding separation of pure fissile materials.

Potential pyrochemical methods for processing fuels and partitioning actinides and other long-lived fission products include: electrorefining using solid or liquid cathode, reductive extraction between molten salt and metal phases (e.g. with multi-staged extractors), fractional crystallisation, drossing or slagging, and volatilisation. In practice processes based on some combination of these are likely to be required. Several of the significant international pyrochemical process concepts that support partitioning are now outlined:

The ANL IFR process is representative of these operations. Irradiated metallic fuel is electrorefined in LiCl-KCl salt. Uranium is electrotransported to a solid iron cathode where it deposits in purified form. Plutonium, minor actinides (MA) and reactive fission products collect as chlorides in the salt. Unreactive fission products generally accumulate as metallic solids in the anode baskets. When sufficient fuel batches have been treated, the plutonium is recovered as a U-Pu-minor actinide alloy in a liquid cadmium cathode that uses a "pounder" to maximise the electrodeposition of mixed fuel solids (U-Pu-MA) in Cd. When the salt from the electrorefiner is sufficiently loaded with fission products, it is treated by lithium reduction/molten metal extraction using a "pyrocontactor" to separate most of the actinides whilst nearly all of the lanthanides are left in the salt. The partly reduced salt is then contacted with Cd-U alloy to remove the remaining transuranic elements with a minority of lanthanides which are returned to the electrorefiner. Residual fission products are removed from molten salt by ion exchange using zeolite to enable immobilisation as a ceramic HLW form.

In 1990, CRIEPI developed a partitioning concept for HLW from PUREX reprocessing. Arising from this, Rockwell, CRIEPI and Missouri University collaborated in developing a process based on separation of actinide and lanthanide chlorides from noble metals and active metal chlorides by consecutive reductive extractions from chloride salt medium into liquid cadmium. Actinides were then separated from lanthanides in an electrorefiner by electrodeposition of actinides on a cathode immersed in LiCl-KCl electrolyte. This differs from the IFR proposal, and the key part is selective actinide electrorefining (electropartitioning). In the late 1990s, CRIEPI with Rockwell, and Missouri University have been developing a further process based on conversion of: PUREX wastes from nitrates to oxides (heat) and oxides to chlorides (Cl_2 and carbon) followed by dissolution in LiCl-KCl eutectic. Noble metals (Fe, Zr, etc.) are reductively extracted into molten cadmium whilst actinide and lanthanide chlorides are subsequently reductively extracted into liquid Cd. Uranium is electrochemically removed from actinides and lanthanides using a molten salt. Actinides are then separated from lanthanides using multistage countercurrent molten salt/reductive metal medium. CRIEPI initially proposed electropartitioning to separate actinides and lanthanides but have more recently turned toward reductive molten metal/salt countercurrent reductive extraction.

ATW fuel processing and partitioning was conceptualised during 1999 as the proposed system for the USDOE accelerator transmutation of waste. It comprised firstly a modified PUREX solvent extraction process (called UREX) to remove a large throughput of uranium as a US Class C waste (< 100 nCi/g actinides) from spent oxide fuel wastes without separation of Pu from high level wastes. The HLW and actinides are precipitated as oxalates, filtered, calcined to oxide form and fed to an electrometallurgical process, "Pyro A". Technetium and iodine are fed to target fabrication. Secondly Pyro A provides a high separation of actinides from reactive and noble fission products. The oxide waste is reduced to metal form and actinide elements are separated by electropartitioning (selective electrotransport of actinides away from RE in molten chloride salt). Actinides are used to make ATW fuel, and reactive and noble metal fission products are immobilised in ceramic and metal waste forms. Thirdly, a chloride volatility process (called Pyro B) separates and recycles ATW irradiated fuel. Actinides are separated by electropartitioning for ATW fuel fabrication, irradiation and transmutation. Technetium and iodine are separated for target fabrication, irradiation and transmutation. Reactive and noble metal fission products are immobilised in ceramic and metal waste forms respectively. One uncertainty in use of electropartitioning at the industrial scale is the high current density requirements with possible significant over-potentials that reduce separation effectiveness.

In summary, fast flux systems with competitive capital and operating costs and low environmental impacts may be needed in the 21st century as uranium and fossil fuel reserves decline. The high recovery and effective waste management of long-lived radioisotopes may be important to public acceptability within the context of fuel cycle costs that can be supported by the electric utility companies. The integration of simplified pyrochemical processing, partitioning and fuel fabrication methods that meet the needs of an advanced reactor system is important.

Overview of historic BNFL pyrochemical work

BNFL has a considerable history of pyrochemical technology. A research programme was undertaken at Springfields in the UK between 1963 and 1969 to investigate the production of uranium metal by electrowinning in high temperature molten salt systems. At the close of the project a 26 inch cell had been operated for several days with a mean output of 1 kg U/hour and current efficiencies of 60%.

There is also a pyrochemical element in the manufacture of Magnox (uranium metal) fuel on a scale of hundreds of tons per year at the Springfields plant. Uranium tretrafluoride is reduced with magnesium metal in a thermite-type reaction initiated at 650°C and reaching 1 500°C at its maximum.

In the 1990s BNFL's advanced reprocessing programme examined a number of technologies for future fuel treatment and recycle applications. The molten salts project started to develop a capability within the company and established many links internationally. Small-scale experimental programmes were established with a number of assessments and pre-conceptual design studies. Molten salt technology was assessed as part of a fluoride volatility programme. Work, much of it now in the open literature, was also performed on the dissolution of uranium dioxide in carbonate melts. Laboratory scale studies were also performed on graphite corrosion by aggressive metal melts.

In 1997, it was decided that BNFL undertake work on the treatment and immobilisation of molten salt wastes. Practical laboratory work explored options for removing inactive fission product simulants from the salt and their immobilisation in a durable waste form. Also in 1997 Magnox Generation, who in 1998 were integrated within BNFL, posted a secondee to Argonne National Laboratory. This secondment resulted in a study of the technology for application to BNFL's Magnox fuels. BNFL's R&T department has a team dedicated to the support of its existing vitrification facilities, and towards the development of future immobilised waste forms. Members of this team are involved in the molten salts development programme. BNFL has also developed powder handling technologies including mills and classifiers for application in active glove box environments. It is believed that this technology will be transferable to treatment steps within pyrochemical processes.

In 1999 the molten salts programme came under re-assessment. The project is now directly funded from Thorp business unit. The project has passed through BNFL's innovation process, and as a result is well targeted, structured on risk management and the resolution of parameters key to the successful final application of the technology (showstoppers). The project now has a broad remit, looking at all of the major aspects of the technology, engineering and industrialisation of the process, environmental and safety implications. There is a keenness to extend international collaborations between BNFL and other organisations towards the eventual successful application of pyrochemical technology in the most suitable commercial context.

BNFL's new technology centre, known as the BTC, is under construction and commissioning. It is anticipated that in future years much of BNFL's pyrochemical work will be performed in this facility (see Figure 1).

BNFL electrorefiner programme

Programme

BNFL Thorp has started a limited experimental R&D programme in the field of molten salt technology to recycle irradiated fuel, partition long-lived radioactive waste and condition it for disposal. A significant part of this work is the design, construction and testing of a chloride salt electrorefiner which will be used to determine various kinetic and thermodynamic data as well as to explore and develop further process features. The planned duration of the test programme including commissioning of the glove box, electrorefiner and other test equipment is three years. The main parameters of interest are: electrodeposition rates, distribution coefficients between solid metallic and molten salt phases, separation factors between molten salt and liquid metal phases, and gravimetric recoveries. These are expected to depend principally on composition, voltage (current density), geometry, temperature and mixing.

Tests will include separate stages such as electrodepositing pure uranium (solid cathode), plutonium with uranium (liquid cadmium cathode), and uranium with plutonium and minor actinides rejecting simulated lanthanide fission products (electropartitioning at solid or liquid cathode, or reductive extraction with molten salt transport). To save time in gaining as much useful data as possible, simulated intermediate feeds (for example, pre-enriched in plutonium to avoid having to deposit surplus uranium) will be used where appropriate. The salt medium will usually be the lithium-potassium chloride eutectic under a pure argon atmosphere in a dedicated alpha-active glove box system.

The stages of the experimental programme are:

1. Equipment commissioning and salt preparation.

2. Pure uranium electrotransport.

3. U-Pu-Am electrorefining.

4. U-Pu-Am-Np-Cm electrorefining.

5. U-Pu-Am-Np-Cm and lanthanides electrorefining and electropartitioning.

6. Li/Cd reductive extraction of U-Pu-Am and lanthanides.

7. Electropartitioning actinides and lanthanides with a liquid bismuth cathode.

Where plutonium is allowed to accumulate in a given volume of salt from several successive batches of fuel (semi-continuous processing), various factors could limit the number so treated. Which factor is limiting depends on the particular fuel and other conditions. In thermodynamic terms, the Pu:U ratio should not exceed about 10^3:1, or the electrorefined uranium at the solid cathode may suffer contamination by Pu. Practically, owing to polarisation at high current densities, the Pu:U ratio may need to be as small as 3:1 or possibly up to 13:1. The uranium trichloride concentration should not fall below around 0.4 mol % depending on the required electrotransport rate and polarisation overpotential with a risk of co-deposition. $CdCl_2$ may be added during runs to restore dissolved UCl_3 concentrations. Above 2 mol %, UCl_3 is reported to polymerise so that electrotransport kinetics do not improve further.

ANL has utilised a design value of 0.07 A cm^{-2} current density with a normal operating voltage range 0.2 V to 0.4 V area in the industrial-scale electrorefiners for low enriched fuel. This corresponds to around 0.2 V overpotential at 2 mol % UCl_3 uranium in salt. A substantial increase in current density is highly desirable, but likely to be prevented by various factors of which a quantitative understanding is sought.

Facility

BNFL has designed an experimental electrorefiner with rotating anode basket for metallic "fuel", rotating solid steel cathode, static liquid cadmium cathode, cadmium pool and molten salt stirrer with paddle features. The electrorefiner components, including the containing vessel, are generally fabricated in low carbon steel. The vessel has an ID of 96 mm and internal height of 336 mm. The removable vessel head is fitted with a centrally mounted stirrer with hollow shaft that enables sampling of the Cd pool, and four radially situated ports. Three are large and dedicated as a group (although interchangeable in use) to the fuel basket anode, iron cathode and liquid cadmium cathode. It is possible, for example, to maximise electrotransport by using a single anode, dual cathode mode of operation. The smaller port is for general use including placing of the reference electrode, e.g. Ag/Ag$^+$ type, and sampling of salt and Cd (a 6 mm sampling tube takes about 5 ml).

Sampling is planned during runs of both the liquid cadmium cathode and the cadmium pool. The sample size to be withdrawn from the liquid cadmium cathode is expected to be around 1 ml in volume and may be greater for other liquids. A molten salt volume of ~0.8 litre and a $PuCl_3$ concentration of ~150 g/L gives a maximum desired inventory of around 100 g of plutonium. The anode basket is of cylindrical shape and has a fine mesh within a coarse basket. A solid iron cathode mandrel for uranium electrodeposition, 16 mm in diameter and 40 mm long used, is fitted with a non-conductive alumina end-cap disc to improve uniformity of voltage distribution and prevent downward growth of uranium that would cause early short-circuiting.

Engineering drawings have been produced for test equipment and fabrication is under way. An existing glove box is being modified to include two heated wells. In one fits the electrorefiner, while the other will be used for experiments on extracting molten salt with reductive metal alloys, as well as chemical preparation, analysis and other purposes. Design, equipment fabrication and installation are intended to be completed by the end March 2000 with commissioning starting in April.

Summary

BNFL has instigated an experimental programme on molten salt process technology embracing electrorefining, electropartitioning and reductive extraction as applied to simulated metallic fuel. Starting from simple systems and moving to more complex ones, it will cover separations of uranium, plutonium, americium, neptunium, curium and selected lanthanides in a three-year experimental programme to be completed by April 2003.

Further experimental studies

The project has been divided into packages of work reflecting the process stages likely to be needed in an industrial process. Within the resource team, expertise is established in the process stages and has been applied to defining the development programme so as to limit the risks inherent in technology deployment.

If oxide fuels are to be treated by pyrochemical technology based upon electrorefining of metal feeds, as in the ANL process, they must first be converted to a metal form. Paper studies have targeted the issues associated with the oxide to metal reduction, both from a fundamental chemistry and an engineering perspective. BNFL is establishing a development programme to address the issues associated with this oxide reduction step.

Electrorefining is likely to be a key process step in any pyrochemical process targeted at the separation and purification of spent fuels or streams derived from spent fuel treatment. BNFL's intent is to build its current practical skills base in molten salts electrorefining technology. In order to do this, BNFL is participating in the EC 5th Framework Programme on pyrochemistry. As described above, BNFL engineers have designed an electrorefiner suitable for operation within a plutonium active glove box.

BNFL has collaborated with several international groups by secondment and contract research. Practical electrochemical measurements in nitride systems are under way. A contract placed at a US laboratory examined the pyrochemical separation of uranium from fuel cladding materials. This work has concentrated on the decontamination factors of the uranium from the cladding material and vice versa, and how the electrochemical dissolution of the cladding material affects the behaviour of the melt system. Other factors likely to play key roles in the operation of plant include the behaviour of anodic sludges and of secondary volatiles.

Crucibles of various types are used throughout pyrochemical processes to contain both molten salts and aggressive molten metals. In order to ensure the maximum lifetime of these crucibles and hence to reduce the quantities of secondary wastes generated, BNFL is developing a number of crucible coating materials.

BNFL believes that the decontamination of the salt solvent from fission products, and the recycle of this salt solvent, will be crucial to the economics of any process. Such recycle should minimise the volume of HA waste generated, that will occupy expensive storage and repository space, whilst optimising the necessary durability of the waste form. The alkali metal and chloride ions are particularly mobile, and can often be incorporated within wastes only at low percentages.

A number of studies are ongoing to examine the separation of fission products from the salt. Extensive studies have now been completed by BNFL on the treatment of various simulant salt wastes with phosphate precipitants. It is the intention to explore zeolite and other ion exchange materials in the future. BNFL's programme has examined the incorporation of inactive waste simulants, and has attempted to optimise the waste loadings. A number of waste forms including glasses, glass-ceramics and ceramics are considered.

BNFL has a capability in thermodynamic modelling of molten salt systems. Understanding the thermodynamic properties of the systems will be crucial to the successful application of the technology.

Engineering studies

BNFL believes that as a commercial company with a proven track record in the design, construction and both safe and economic operation of commercial scale plants, it has considerable experience to offer in industrialising pyrochemical processes. Within the current molten salts project there are a number of engineering projects. Any application of pyrochemical technology to irradiated nuclear fuel will require a head end process to prepare for the subsequent treatment steps. Head end processes used for aqueous systems have so far been generally assumed to be also optimal for pyrochemical reprocessing. BNFL does not consider this to be necessarily so. Findings are that there are potential processes other than the bulk shear which offer advantages when coupled to a pyrochemical system. Development issues for the head end have been identified.

BNFL is currently performing two internal first pass engineering design studies, examining the issues of applying pyrochemical technologies to the treatment of spent nuclear fuels. These are broad cost studies, examining the cost obstacles for process intensification. Consideration is being given to the scale-up and industrialisation of molten salts, and the implications of scale-up on how the process is likely to be performed. A reference base will thus be established for engineering and costs associated with a particular fuel type. Internationally developed molten salts processes are generally based upon batch processes, and the electrorefiner is usually a "pot" process. BNFL is also performing a radical design study, outlining the generic steps that any pyrochemical process requires, defining the current engineering practice applied to these processes and attempting to establish a radically engineered flowsheet. This is aimed at increasing throughputs, minimising mechanical operations and increasing projected process reliability.

In 1995, BNFL performed a design comparison of two process options, RIAR oxide electrorefining and ANL lithium reduction with metal electrorefining, for a LWR processing complex. The study involved Sellafield and Risley staff and was based upon a reprocessing plant of 150 t(HM)/a together with its principal satellite plants, serving five LWR reactors as part of a reactor island complex. Process steps were examined and costed.

In 1998, BNFL reviewed the concept and design of a molten salts waste treatment facility for RP2 which had been performed by Toshiba. BNFL then performed an alternative design of facility that in both cases was compared with an equivalent radical PUREX facility. Waste loadings and formulations using zeolite, phosphate and silica materials were examined and capital and operating costs assessed.

Also in 1998, BNFL commissioned a cost study in the US that examined the best publicly available information on electrometallurgical (EM) treatment technology for management of gas-cooled metallic irradiated fuel. The study indicated feasibility of EM treatment and developed cost and schedule information underpinned by reference process and facility pre-conceptual designs. Information was developed using a bottom-up generic work breakdown system structured by cost element. Pre-operational, operational and decommissioning costs were estimated to obtain an overall estimated life cycle cost. Statistical modelling of cost probability distributions was applied.

General and specific design requirements were developed. The design philosophy included safety, operability and HAZOP studies. Process option studies were performed and the logic/justification to select the reference EM treatment process design was formalised. An industrial engineering study was conducted to confirm that the specified throughput and required maintainability could be achieved using process cells and other areas as described. A facility design concept was developed including 3-D drawings. This design included a pre-conceptual layout of major systems, structures and components. It was found that EM treatment is technically effective for separating and immobilising the bulk of radioactive species from irradiated thermal metallic fuel. High-level waste is immobilised as a ceramic waste form and uranium and cladding waste streams are produced in compact metallic forms for intermediate storage.

International links

Development of any major new nuclear process is probably beyond the means of any single commercial organisation. Throughout the last ten years BNFL has been active in establishing links with international organisations involved in pyrochemical technology. There is a keenness that both these and new interactions are nurtured, to encourage appropriately focused commercial development and application of pyrochemical technology.

Pyrochemical programme interaction between Japanese organisations and BNFL has recently included secondment of BNFL personnel, most recently concentrating on the measurement of fundamental thermodynamic properties. A BNFL engineer has been granted an international fellowship to JNC, Tokai. It is intended that as part of these fellowship studies, work will be performed to address the industrialisation of pyrochemical processes. BNFL has performed joint pyrochemical plant design studies with Toshiba, and has current design contracts with other Japanese organisations.

Links between BNFL and US have been active in the past, with BNFL secondees on separate projects to Argonne National Laboratory and Babcock and Wilcox. A number of links have been made with US universities. BNFL's subsidiary companies BNFL Inc. and Westinghouse Nuclear Services provide additional technical and commercial links and programme support.

BNFL has also had a number of visits, collaborative projects and work packages with RIAR Dimitrovgrad.

BNFL is part of the European Commission 5th Framework Programme on partitioning and transmutation, with collaborative partners including CEA – Marcoule, CIEMAT – Madrid, NRI – Řež Czech Republic, ENEA – Casaccia Italy, ITU CRIEPI Karlsruhe. BNFL is contracting a portion of its work to AEA-T Harwell, who have extensive links with BNFL's research programmes and a proven history of working together with BNFL.

BNFL uses fundamental thermodynamic information obtained through links and contracts with the UK National Physical Laboratory. There is a hope to extend collaborations with both French and Japanese researchers within this subject.

Within the UK, BNFL also has contracts and good working relationships with universities active in pyrochemical technology development.

BNFL pyrochemical strategy

BNFL is a commercial company operating to provide a safe world class service to its customers whilst earning a return to its shareholder, the UK government. As a commercial company BNFL expects there to be a future financial return on investments made into research and development. BNFL maintains a core research base to support existing plant operations, to advance future technology developments and to provide a strategic function.

BNFL's molten salts project operates within the company's commercial environment and is subject to BNFL's innovation process management system. As such the project has clearly defined criteria and objectives upon which the internal customer demands delivery.

The potential of pyrochemical technology to the nuclear fuel cycle is recognised within BNFL, as are the advances to be made to allow development and application of the technology internationally. There is also an understanding that a single organisation will find it very difficult to develop and apply the technology independently at an economic commercial scale.

Investment in BNFL's molten salts programme will be focused on areas where BNFL can "add value" to international pyrochemical development programmes. The project is being applied to develop the technology in international partnership against clear market opportunities. These market opportunities are generically considered to include: fuel conditioning to minimise HA waste volumes, treatment of unstable fuels, recycle of fast reactor and MOX fuels and potential options for partitioning and transmutation recycle.

BNFL supports the investigation of the potential benefits of molten salt technologies to support future nuclear fuel cycle development.

Conclusions

- BNFL currently fabricates metallic uranium fuel at an industrial scale using pyrochemical processing including fluorine gas production, and also has a background of pyrochemical development programmes that include electrorefining, fluoride volatility and low temperature carbonate salts.

- BNFL has a pyrochemical R&D programme that involves collaboration with international partners and emphasises industrialisation of processes and design of plant. We see the expertise needed for partitioning as important for all types of pyrochemical spent fuel treatment and regard pyrochemical treatment as being a leading contender for potential partitioning and waste management of long-lived radionuclides.

- Industrialisation studies are being performed on ANL and RIAR type processes and fundamental chemical studies being made in key areas.

- A molten chloride salt electrorefiner is under construction with commissioning commencing in April 2000. It will be used in the collaborative EU 5th Framework Programme concerning partitioning of minor actinides from fuel or high level waste to investigate electrorefining of simulant fuel containing up to 100 g Pu and selected minor actinides and lanthanide elements.

- This programme is business driven and targets process and plant uncertainties so as to reduce the risks of process and plant industrialisation.

- The aim is to develop capability, possibly in partnership with international collaborators, to meet future market needs for partitioning or fuel processing purposes, if commercially justified.

- Novel processes and reaction media are being examined and patented. These are considered fundamental to further simplification of molten salt technology to reduce process and production risks and increase the likelihood of technology deployment.

Acknowledgements

The authors would like to acknowledge the expertise and involvement within the BNFL molten salt programme of the following: Peter Wylie, Nigel Donaldson, Richard J. Taylor, Phil Mayhew, Sean Morgan, Keith Franklin, Paul Gilchrist, John Cogan, Justine Hatter, Paul Sculley, Andrew Nairn.

Figure 1. Uranium active rig hall BTC

ASSESSMENT OF PYROCHEMICAL PROCESSES FOR SEPARATION/ TRANSMUTATION STRATEGIES: PROPOSED AREAS OF RESEARCH*

Bernard Boullis, Philippe Brossard
CEA/DCC
B.P. 171, 30207 Bagnols-sur-Cèze, France

Abstract

Pyrochemical processes involving high-temperature separation in molten salt media could represent a promising alternative to the hydrometallurgical separation processes adopted as the "reference route" to assess separation and transmutation strategies for long-lived radionuclides within the scope of the first research area specified in the French radioactive waste management law passed on 30 December 1991. Such processes have a number of known advantages, for example their inherent compactness. Moreover, the range of potential options regarding transmutation (and the fuels and targets associated with the scenarios under consideration) provides additional impetus for investigating pyrochemical routes, notably in view of the anticipated limits and problems inherent in aqueous processes (solubilisation of highly irradiated and/or relatively hot objects and radiation sensitivity of process reactants in particular).

Various concepts for pyrochemical reprocessing techniques have been studied, developed and tested (in some cases at a pilot scale) throughout the world. The generic approach consists in dissolution of spent fuel elements in a molten salt eutectic at a temperature of the order of 500-800°C, followed by some degree of selective recovery of constituent elements for recycling or for conditioning. Several media and a variety of techniques have been – and continue to be – investigated: chloride or fluoride media, separation by extraction (using molten metals), by electrodeposition or by precipitation. The IFR concept of electrorefining in a chloride medium developed at the Argonne National Laboratory (USA) is no doubt the most mature of these processes and has been tested experimentally to recondition several hundred kilograms of spent metallic fuel.

The primary objective of the proposed research programme is to provide the most detailed possible background data for the required 2006 assessment, not only on specific applications identified at the present time (e.g. processing of transmutation targets for multiple recycling) but also from a broader perspective extending to applications that are conceivable but not yet sufficiently defined or to relatively remote prospects (such as dedicated fuels for advanced concepts). The purpose is thus to conciliate various requirements and to acquire information of a sufficiently generic nature that it would be suitable for implementation in applications that could emerge or be specified in the coming years.

---

* The full text of this paper is unavailable.

For the immediate future, the project involves three major stages to which the following time frame has been assigned:

- 1999–2002: Exploratory studies of various operations covering several potential applications with metals or ceramics, for the purpose of acquiring the data necessary for a preliminary selection of process concepts (media, technological requirements, etc.).

- 2002–2005: Applied research on selected concepts, involving radioactive experimentation at laboratory scale on simulated products ("reconstituted" specimens).

- 2005 and beyond: Demonstrative experiment(s) at laboratory scale on actual representative specimens (notably targets irradiated in the PHENIX reactor).

An important aspect of this programme will involve fundamental research on process chemistry and engineering. This work is indispensable to provide a firm basis for selecting the concepts and subsequently for controlling and adapting them as necessary to meet the changing requirements of specific applications.

The broad research programmes require a variety of equipment, ranging from "cold" laboratories to shielded cell lines for spent fuel experimentation. The facilities now operational or soon to be commissioned in the ATLANTE complex at Marcoule will, with some modifications, provide the CEA with most of the material resources necessary to complete the planned programme. Moreover, these research activities are proposed within a largely open context – not only for the fundamental aspects (with a major contribution from the French scientific community, notably under the PRACTIS Research Group) but also for overall strategy studies (GEDEON Research Group, assessments performed under the European 5th PCRD) and more application-oriented research (co-operation possible or desirable with foreign organisations having experience and/or similar objectives in these areas).

SESSION II

Role and Requirements
of Pyrochemical Separations
in the Future Fuel Cycles

Chairs: P. Wilson, M. Hugon

ROLE OF PYROCHEMISTRY IN ADVANCED NUCLEAR ENERGY GENERATING SYSTEM

L. Koch
European Commission, Joint Research Centre
Institute for Transuranium Elements, Nuclear Chemistry
Postfach 23 40, D-76125 Karlsruhe, Germany

Abstract

The on-site dry reprocessing of spent nuclear fuels offers advantages versus the present method of aqueous reprocessing. In case of power stations with fast neutron spectra (including accelerator-driven systems) the purification needs of fissile materials are less but depend on the fuel matrices used. The limits of ecological advantages by partitioning of long-living radioactive nuclides are given. Based on the present state of knowledge and, compared to future requirements, the economic and social (proliferation) aspects are addressed.

Scope

The present low price of uranium makes recycling of plutonium and uranium economically unattractive. Nevertheless, in the future, uranium resources will become exhausted. If there is no reasonable alternative energy generation option, nuclear power will have to be extended to TRU element recycling and the intermediately stored fuel will not be disposed of in a geological repository.

There are two main scenarios:

- The transition from light water reactors (LWR) to fast breeder reactors (FBR) where not only uranium and plutonium are recycled, but also the self-generated minor actinides (MA) (Np, Am, Cm) together with some long-living radiotoxic fission and activation products. This option is being considered in France, Japan, China, South Korea and the Russian Federation. Aqueous as well as pyrochemical partitioning processes are being considered. The fuel consists mainly of mixed oxides. CRIEPI of Japan is developing a metallic fuel (Figure 1).

- The transition of the uranium fuelled LWR to a Pu (MOX) fuelled LWR and/or FBR with a separate fuel cycle to transmute the long-lived radiotoxic minor actinides as well as fission and activation products (double strata). This option is considered in the ATW of LANL, in the Omega project by JAERI and in the ADS of the Rubbia concept. All approaches are based on the use of pyrochemical partitioning processes. Again, the final matrix consists mainly of mixed oxides, but of different even non-fertile nuclides [1] (Figure 2).

The aqueous reprocessing of nuclear fuels by the PUREX process can be extended to recover Np. New processes partition minor actinides from the high level waste [2-5]. The aqueous processes have been developed and their technical feasibility demonstrated [6], but they are limited by the radiolytic degradation of the extractant caused by short cooled fuels and high minor actinide concentrations. Consequently, spent fuels would need a cooling time of several years and diluted minor actinide streams. The aqueous processes compared to the pyrochemical partitioning process described below have to be more voluminous, due to the low critical mass of ^{242m}Am (19 g). In fast neutron spectra, the neutron poisoning by fission products becomes minimal. Hence the pyrochemical processes with their lower fission product decontamination is acceptable.

Pyrochemical processes

Several molten salt partitioning concepts have been investigated, mainly by electrochemistry, possibly supplemented by counter-current extractions [7] or fractionated precipitation [8,9]. Although those using molybdates [10] and tungstates [11] avoid the use of corrosive halogens, they need higher process temperatures (see Table 1). For these reasons they are less favourable than partitioning in molten halides. Since the Gibbs' energies of minor actinides from the lanthanides are more different for chlorides than for fluorides (Figure 3), the latter are not considered – except in a molten fluoride fuel concept, as once used (Table 1) for the molten salt breeder (at ORNL, US). This is due to the long-living ^{36}Cl activation product of 3×10^5.

The electro-potential of actinides in molten alkali chlorides permits a separation of metals, and of uranium and neptunium oxides, but not of the higher actinide oxides (Figure 3). Therefore, from the two proposed processes, the DOVITA process separates only UO_2 and NpO_2 by electrolysis, PuO_2 is precipitated from the melt and americium and curium are presently fractionated precipitated [8,9].

On the contrary, the metal refining as developed by CRIEPI [12,13] separates all transuranium elements (Figure 3). There seems to be a difficulty for americium, which tends to form $AmCl_2$ during electrolysis. Studies underway try to circumvent this effect [14]. Since the reprocessing of spent fuel by pyrochemistry has already been demonstrated for short-cooled fuel [15], and since the (fast neutron) critical mass of [241m]Am is 3.3 kg, a compact on-site facility seems possible, integrated in a single fast reactor or a nuclear park.

What are the advantages?

Economics

The pyroreprocessing directly after discharge of the spent FBR fuel shortens the out-of-pile time of the fissile inventory and thus reduces its doubling time. Moreover the minor out-of-pile decay of [241]Pu limits the loss of fissile material and avoids the build-up of the radiotoxic [241]Am. The technical feasibility of the on-site partitioning of actinides has not been fully demonstrated. The present reprocessing of the spent EBR II fuel separates only uranium and conditions the transuranium elements together with the fission products for waste disposal [15]. Nevertheless, it is estimated that a pyrochemical partitioning due to its compactness is superior to aqueous reprocessing. The separation yields for the two are the same [16] (Table 2). Nowadays, the technical feasibility studies for the aqueous processes, however, are by far more advanced than those for the pyrochemical ones.

For the partitioning of spent fuel of accelerator-driven fast reactors, a separate fuel cycle has to be established with a decontamination of the two minor actinides Am and Cm from the lanthanides to be ten times better than for the spent homogeneous minor actinide containing FBR fuel. This is due to the high minor actinide content of about 20% in the waste burner fuel, which cannot tolerate a lanthanides fraction similar to that in the 2% MA FBR fuel [17]. Consequently, the partitioning effort for this fuel is twice that for the FBR fuel.

The proposed fuels will be mainly either Zr based alloys, mixed oxides or nitrides. If it is decided to use oxide, one has – in the case of metal electrorefining – to convert oxide in metal or chloride and metal into oxide, which would considerably complicate the process and enhance the cost. Suitable processes are being developed [18]. Remote fabrication is required for all minor actinide-containing fuels. The vibro-compaction of milled oxides obtained by the DOVITA process has been successfully demonstrated [19]. Injection casting is being used for U-Pu-Ir alloys. Due to the high vapour pressure of Am powder metallurgy has been suggested recently. Only Np and Pu separated by PUREX could be handled by the present glove box technology, which is an advantage for the double strata concept.

Ecology

For the recycling of Pu in LWRs – as is still foreseen in some of the double strata strategies – one has to stress the disadvantage that it would increase the radiotoxicity of the spent fuel considerably, because one-third of the plutonium will be transformed into americium and curium.

The conversion of ceramic fuels into metals and vice versa for a metal refining will increase the losses and consequently reduce the P&T effort.

On-site reprocessing, as pointed out above, reduces [241]Am build-up and hence its transformation to curium nuclides in a fuel cycle.

At present, the transport of spent fuel is regarded as being a great public hazard [21]. On-site reprocessing would eliminate the need to transport spent fuel for a centralised reprocessing and instead, one could cool the solidified waste – purified largely from its long-lived radiotoxicity – at the reactor site before it would be sent directly to a geological repository.

Social acceptability

The latter argument will increase the social acceptance of on-site reprocessing. The re-use of the self-generated minor actinides in a FBR will increase the energy output by 10% in the waste burner concept. The radiotoxicity inventory has to be collected from different power stations and burned at selected sites, where it will meet very low local public acceptance.

The pyroreprocessing, and especially the electrorefining in a liquid Cd cathode, deposits the TRU elements together. Such a mixture is not suitable for a nuclear explosive. Moreover, the inherent incomplete separation from fission products (compared to aqueous reprocessing) requires remote handling of the TRU elements. Altogether, this increases the proliferation resistance of a pyrochemical on-site reprocessing compared to PUREX produced Pu and Np in the double strata concept.

Conclusion

- The advantages of pyrochemical partitioning: compactness and processing of short cooled spent fuel shows up best in an on-site P&T of an integral reactor concept.

- From the present stage of knowledge, only the metal refining in molten alkali-chlorides seems technically feasible.

- The use of metallic fuels is more reasonable because no conversion of ceramics into metals and vice versa is needed.

- However, the incompatibility of such a fuel with Pb or Pb/Bi coolant may require gas-cooling for future advanced nuclear energy generating systems.

- The homogeneous recycle of MA in future FBR fuels requires a remote handling, which enhances the proliferation resistance in all parts of the fuel cycle.

- The same proliferation resistance does not exist for a double strata scheme with the separated Pu. The cost advantage for having only the MA concentrated in a separate fuel cycle with remote handling has to be balanced with the extra effort to develop a separate fuel cycle and the accelerator technology needed.

- The higher MA concentration in waste burner fuels requires a ten times better MA/lanthanide separation.

In essence, from the state of the art of pyrochemical partitioning, the metallic fuel cycle is the better choice.

REFERENCES

[1]   R.J.M. Konings, R. Conrad, G. Dassel, B. Pijlgroms, J. Somers, E. Toscano, "The EFTTRA-T4 Experiment on Americium Transmutation", EUR Report, in press.

[2]   E.P. Horwitz, D.G. Kalina, H. Diamond, G. Vendegrift, W.W. Schulz, *Solvent Extr. Ion Exch.* 3, 1985, 75.

[3]   J.-Pr. Glatz, C. Song, X. He, H. Bokelund, L. Koch, "Partitioning of Actinides from HAW in a Continuous Process by Centrifugal Extractors", Proc. of the Special Symposium on Emerging Technologies in Hazardous Waste Management, 27-29 September 1993, Atlanta, Georgia, D.W. Tedder, ed., ACS, Washington, DC.

[4]   M. Kubota, I. Yamaguchi, Y. Morita, Y. Kondo, K. Shirahashi, I. Yamagishi, T. Fujiwara, "Development of a Partitioning Process for the Management of High-Level Waste", Proc. of GLOBAL'93, p. 588 (1993).

[5]   C. Madic, P. Blanc, N. Condamines, P. Baron, L. Berthon, C. Nicol, . Pozo, M. Lecomte, M. Philippe, M. Masson, C. Hequet, M.J. Hudson, "Actinide Partitioning from HLLW using the Diamex Process", Proc. of the Fourth Int. Conf. on Nuclear. Fuel Reprocessing and Waste Management, RECOD'94, London, UK, 24-28 April 1994.

[6]   O. Courson, R. Malmbeck, G. Pagliosa, K. Römer, B. Sätmark, J.-P. Glatz, "Separation of Minor Actinides from a Genuine An/Ln Fraction", Proceedings Euradwaste'99, Radioactive Waste Management Strategies and Issues 15, Luxembourg, 18 November 1999.

[7]   K. Kinoshita, T. Inoue, S.P. Fusselman, D.L. Grimmet, J.J. Roy, C.L. Krueger, C.R. Nabelek, T.S. Storvick, "Separation of Uranium and Transuranic Elements from Rare Earth Elements by Means of Mutistage Extraction in LiCl-KCl/Bi System", *J. Nucl. Sci. Technol.*, 36 (2), 189 (1999).

[8]   A.V. Bychkov, S.K. Vavilov, P.T. Porodnov, A.K. Pravdin, G.P. Popkov, K. Suzuko, Y. Shoji, T. Kobayashi, "Pyroelectrochemical Reprocessing of Irradiated FBR MOX Fuel", Proc. of GLOBAL'97, 2, 912.

[9]   A.P. Kirilloovich, A.V. Bychkov, O.V.S. Skiba, L.G. Babikov, Yu.G. Lavrinovich, A.N. Loukinykh, "Safety Analysis of Fuel Cycle Processes Based on "Dry" Pyrochemical Fuel Reprocessing and Vibropac Technology", Proc. of GLOBAL'97, 2, 900.

[10]   O.A. Ustivon (ARRIIM, Moscow, Russia), "Physico-Chemical Validation of Spent U-Pu Oxide Fuel Reprocessing by Recrystallization in Molten Molybdates", Proceedings of Molten Salts in Nuclear Technologies, Dimitrovgrad, RF, 1995.

[11]   V.K. Afonichkin, V.E. Komarov (IHTE, Ekaterinburg, Russia), "Production of Granulated Uranium Oxides by Electrolysis of Tungstate Melts", Proceedings of Molten Salts in Nuclear Technologies, Dimitrovgrad, RF, 1995.

[12]   Y. Sakamura, *et al.*, "Development of the Pyropartitioning Process – Separation of Transuranium Elements from Rare Earth Elements in Molten Chlorides Solution: Electrorefining Experiments and Estimations by using the Thermodynamic Properties", Proc. of GLOBAL'95, 2, 1 185.

[13]   K. Uozumi, Y. Sakamura, K. Kensuke, S.P. Fusselman, C.L. Krueger, "Development of Pyrometallurgical Partitioning Technology of Long-Lived Nuclides", CRIEPI-Report T98011 (1999).

[14]   L. Koch, C. Pernel, T. Koyama, "European Patent Application: Electrorefining of Americium", 00101954.6 (2000).

[15]   J. Laidler (ANL, USA), "Pyrochemical Separation Technologies Envisioned for the US Accelerator Transmutation of Waste System", these proceedings.

[16]   Draft report – "Comparative Study of ADS and FR in Advanced Nuclear Fuel Cycles".

[17]   M. Kurata, A. Sasahara, T. Inoue, M. Betti, J.F. Babelot, J.C. Spirlet, L. Koch, "CRIEPI: Fabrication of U-Pu-Zr Metallic Fuel Containing Minor Actinides", Proc. of GLOBAL'97, Yokohama (1997).

[18]   T. Inoue, T. Usami, M. Kurata (CRIEPI, Japan), J. Jenkins, H. Sims (AEA Technology, UK), "Pyrometallugial Reduction of Unirradiated TRU Oxides by Lithium in a Lithium Chloride Media", these proceedings.

[19]   O.V. Skiba, A.A. Mayorshin, A.V. Bychkov, V.A. Kisly, P.T. Porodnov (REAR, Russia), A.I. Kiryushin, V.A. Rogov, V.Y. Sedakov, (OKBM, Russia), "Nuclear Fuel and Power Complex, Included the Modul-Type Fast Reactor with Low Capacity and the Plant Dry Reprocessing of Fuel and Vibropacked Fuel Pins Assemblies Production", Proceedings, GLOBAL'97, Yokohama (1997).

[20]   C. Thompson, C.C. Wade, "ATW Road Map Report to NEA Expert Group Meeting", Paris, 25-26 Oct. (1999).

[21]   Gordon E. Michaels, ORNL paper presented to Nat. Research Council Stats. Panel, Washington (1992).

Figure 1. Transition of U-fuelled LWR to FBR fuel cycle in a P&T mode

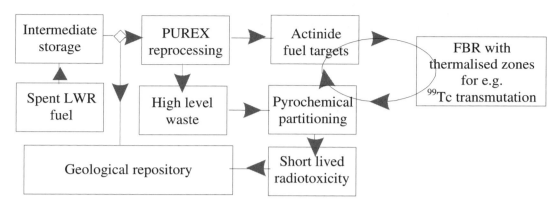

Figure 2. Transition of U-fuelled LWR to U, Pu-MOX fuelled LWR or FBR cycle and a separate P&T cycle for long-living radiotoxic nuclides

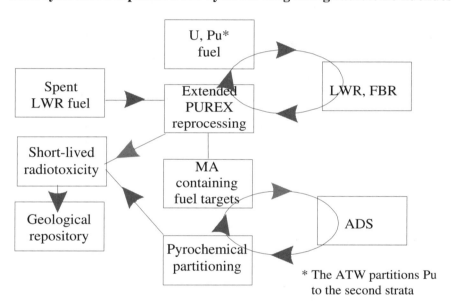

Figure 3. Electropotential relative to Cl₂/2Cl for metal and oxide actinides in molten LiCl-2KCl (data for NpO₂, Cm-Am estimated from literature, other data from A.V. Bychkov, *et al.*, GLOBAL'93)

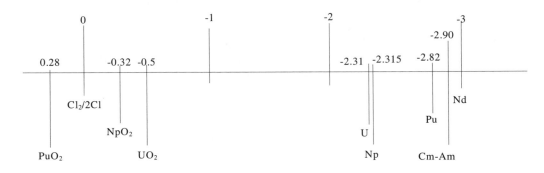

Figure 4. Pyrochemical partitioning processes

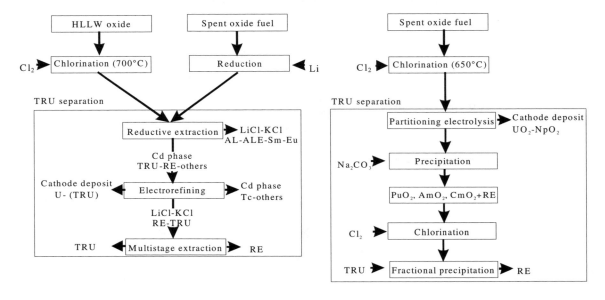

Table 1. Gibbs' energies for fluorides and chlorides

	kJ/eg. F^-/Cl^-	$T°$
Am/AmF$_3$	-460	980
Nd/NdF$_3$	-470	980
Am/AmCl$_2$	-275	680
Gd/GdCl$_3$	-290	680

Table 2. Reprocessing process options. Recovery yield of separated elements/nuclides.

		Standard PUREX	Improved PUREX	Extended PUREX	4 group separation (estimation)[a]	Pyrochemical (estimation)[a]
U		99.9%	99.9%	99.9%		99.9%
Pu		99.0 ~ 99.9%	99.9%	99.9%	>99.85%[b]	99.9%
Minor actinides	Neptunium	60 ~ 95%[c]	99.9%[d]	99.9%	>99.85%	99.9%
	Am	–	–	99.9%	>99.97%	99.9%
	Cm	–	–	99.9%	>99.97%	99.9%
	Lanthanide/MA[e]	–	–	<25%		<25%
Fission products	^{99}Tc	–[f]	80~90%[g]	80~90%	~95%	~95%
	^{129}I	95%	95%	95%		95%[h]
	^{79}Se	–	–	–		–[k]
	^{90}Sr	–	–	–	>99.9%	–[k]
	^{135,137}Cs	–	99.9%	99.9%	>99.9%	–[k]

PYROCHEMICAL REPROCESSING TECHNOLOGY AND MOLTEN SALT TRANSMUTATION REACTOR SYSTEMS

Jan Uhlir
Nuclear Research Institute Řež plc
CZ-250 68 Řež, Czech Republic
E-mail: uhl@nri.cz

Abstract

The main present interest in the development of pyrochemical and pyrometallurgical partitioning methods is caused mainly by the problems in separation of minor actinides from the spent fuel by hydrometallurgical separation methods. However, the dominant purpose of the pyrochemical separation methods development should be associated with the development of new liquid fuel (molten salt) based reactor transmutation systems. These reactor systems should be directly connected with the continuous or semi-continuous separation process ensuring practically perfect exploitation of the reactor transmutation power. Demonstration of the ability of the pyrochemical and separation technologies to meet the demands made upon their fuel cycle is a condition for the realisation of these reactor systems. The experience obtained with the pyrochemical technologies both in the nuclear and non-nuclear spheres until now indicates that satisfying these demands by pyrochemical methods could be feasible.

The present interest in the development of pyrochemical or pyrometallurgical methods of spent fuel reprocessing, the objective of which is the separation of plutonium and minor actinides for their subsequent transmutation in thermal or fast reactor systems to short-lived or even stable isotopes, is caused mainly by the inability of the hydrometallurgical separation methods to comply with all the demands placed upon the separation process for this fuel cycle.

The problems in separation of individual minor actinides and some fission products, which could be solved effectively only by using rather complicated technologies, are particularly concerned.

Some of these drawbacks could be eliminated by using pyrometallurgical processes. High radiation resistance of inorganic reagents allowing faster fuel recycling, considerable process compactness and production of a small amount of waste only could be expected. However, the application of pyrometallurgical processes calls for tackling extraordinary demands on the construction materials and process control, as these processes are proceeding often at temperatures above 500°C under the application of very aggressive chemical reagents.

At present, a technologically feasible conception of spent fuel transmutation seems to be based on a gradual process of plutonium burning and minor actinides transmutation in several steps – at first of the MOX fuel in light water thermal reactors, then in a form of metallic or ceramic fuel in fast reactors and finally in accelerator-driven subcritical systems with liquid fuel. However, each of these reactor systems has different, in some respects very specific, requirements on the spent fuel reprocessing. Some of the requirements may be better satisfied by using hydrometallurgical processes, others by pyrochemical or pyrometallurgical processes. These specific requirements of the individual reactor systems are to be clearly defined to facilitate the development of a reprocessing technology for particular reactor systems. In addition, in the case of the development of pyrochemical and pyrometallurgical methods we have to compare our objectives and results attained with the results of hydrometallurgical procedures already obtained and thus concentrate first of all on areas where the purposeful supplementation or at least partial substitution of hydrometallurgical technologies could be presumed or on areas where the hydrometallurgical procedures and organic agents could probably not be used at all.

Reprocessing of uranium and uranium-plutonium (MOX) spent fuel after several years of cooling by using the hydrometallurgical process PUREX has worked out very well for uranium, plutonium and relatively well for neptunium, too. However, americium and curium could not be practically separated from fission products by this method and that is why other hydrometallurgical methods for separating these minor actinides are under development. Some of the processes like DIAMEX, SESAME, TRPO and others seem to be promising, but their application brings about further extension of hydrometallurgical technology and, naturally, further increases the amount of radioactive waste. The newly developed pyrometallurgical procedures for separation of these minor actinides based on the extraction into liquid metals, on electroseparation from chloride or fluoride media or on the reduction with alkali metals from nitride media have a possibility of being utilised in combination with the PUREX hydrometallurgical process.

A combination of hydrometallurgical methods with the pyrochemical ones utilising the specific merits of both systems may turn out to be advantageous for the transmutation of Pu and minor actinides in thermal and fast reactors.

However, the dominant purpose of the pyrochemical separation methods development should be associated with the development of new liquid fuel based reactor transmutation systems. These new reactor systems should be directly connected with continuous or semi-continuous separation process ensuring practically perfect exploitation of the reactor transmutation power. A demonstration of the ability of the pyrochemical and pyrometallurgical separation technologies to meet the demands made upon their fuel cycle is a condition for the realisation of these reactor systems. These demands are related both to the actual separation procedures and to the verification or development of necessary construction materials, technological facilities and devices.

One of the main technical requirements in the development of liquid fuel reactor system should be the extremely close co-operation of reactor physicists and reactor engineers with chemists and nuclear chemistry engineers developing the fuel production technology for the reactor, and especially the subsequent technology for the separation of transmuted fuel components from the non-transmuted ones which will be returned into the reactor. It will be most advisable to keep one chemical form of fuel for the entire fuel stream system in developing a compact fuel cycle of the liquid fuel reactor system. If this succeeds, then the demand for a continuous or at least quasi-continuous connection of reactor and separation process fuel streams could be realistically envisaged. This results in a fundamental requirement on the choice of separation methods which could be applied in the reactor fuel preparation and especially in the fuel reprocessing. Electroseparation procedures appear in the foreground of interest as reprocessing methods.

The following main criteria on the liquid fuel need to be fulfilled in order to meet the above requirement on keeping the same basic chemical form of the molten salt transmutation reactor (MSTR) fuel matrix and electrolyte composition for electroseparation of its spent fuel components:

1. The supporting fuel matrix must be radiation resistant. The elements the matrix is formed of must have small neutron absorption cross-sections so that they will not cause undesirable neutron field attenuation in the reactor core owing to some nuclear reactions in which they might take part.

2. The matrix must not contain nuclides which could take part in nuclear reactions under the formation of long-lived radionuclides or such short-lived nuclides that comprise gases or nuclides with large cross-sections (e.g. tritium) to a greater extent in their decay series.

3. Melting point of the matrix should be as low as possible.

4. Partial vapour pressures of components in the operation temperatures range must be low.

5. Viscosity of the matrix should be high enough so as to allow good circulation of the melt and at the same time sufficiently stable within the operation temperature range with regard to the design of pumps and control valves.

6. U, Pu and MA in the first place have to be sufficiently soluble in the matrix and must form true solutions to prevent accumulation of these components at the bottom of various parts of the equipment.

7. Materials compatible with the matrix have to be available for the construction of the reactor, its primary and secondary circuit and electroseparation equipment.

8. Decomposition voltages of the individual components the supporting matrix is composed of have to be higher than the decomposition voltages of the actinide and fission product components.

9. The supporting fuel matrices have to be regenerable to a great extent.

10. Methods for the fixation of radioactive waste from this process must be available.

11. Basic components of the matrix should be from raw materials which are accessible with relative ease.

Most of the above requirements are met relatively well by the alkali metal fluorides (LiF, NaF, KF) and partially by the alkaline-earth metal flurides (BeF_2) (in the case of lithium fluoride the lithium isotope [7]Li is to be considered, because [6]Li isotope would be a significant source of a very undesirable radionuclide – tritium ^{3}H) suggesting that the entire MSTR fuel cycle could be founded on a fluoride basis. This statement is well supported by the experience with a MSR operation and MSBR project in ORNL, USA, experience with the fluoride reprocessing of fuel from the RAPSODIE fast reactor in the Attila facility at Fontenay-aux-Roses in France, a similar Japanese project and experience from the Czech-Russian (respectively former Czechoslovak-Soviet) co-operation on the development of a project for fluoride reprocessing of spent fuel from BOR-60 fast reactor in the Fregat facility [1-4].

All the projects realised in the 60s, 70s and 80s confirmed the functionality of the processes. However, they have also demonstrated great difficulties, especially of a material character. Their present appraisal indicates that they were ahead of their time.

The Czech research programme in the P&T area, also proceeding from this idea, is founded on the conception of a subcritical accelerator-driven MSTR with fluoride salts based liquid fuel, the fuel cycle of which is grounded on pyrochemical or pyrometallurgical fluoride reprocessing of spent fuel. The Czech Republic assumes that this research programme, launched four years ago, will be further advanced within a wide international co-operation [5].

Experimental and theoretical work in the area of development of pyrochemical technologies for ADS is oriented in the Czech Republic mainly towards the following fields:

1. Technological research in the field of "fluoride volatility method", directed at the verification of the suitability of a technology for fast reactor spent fuel reprocessing which may result in a product the form and composition of which might be applicable as a starting material for the production of liquid fluoride fuel for MSTR. Consequently, the objective is the separation of a maximum fraction of the uranium component from Pu, minor actinides and fission products.

2. Research on material and equipment for fluoride salts media connected with the experimental programme on ADETTE technological loops. The objective of the programme is verification of the suitability of SKODA MONICR construction material for fluoride melts, development of selected devices (first of all of pumps) for fluoride melts and acquirement of fluoride melts manipulation in greater amounts (Figure 1).

3. The newly begun laboratory research on electroseparation methods in fluoride melt media in relation to the study of their properties.

Research and development in the area of pyrochemical technologies, concentrated above all in the NRI Řež, is being carried out within the framework of a national programme on transmutation financed by the Ministry of Industry and Trade and by the Radioactive Waste Repository Agency. The co-operating organisations are first of all the SKODA NM and Energovyzkum Brno, the main field of their activities being in the area of materials and equipment.

Figure 1. View of the experimental molten fluoride salt loop ADETTE-0

The Czech national conception in the area of P&T research proceeding from the national energy policy envisages further development of nuclear power industry in the Czech Republic primarily in relation to the nuclear power industry development in most of the EU states. The solution to the problem of the end of fuel cycle in the form of partitioning and transmutation of spent fuel is considered to be real in this conception and technically viable in the relatively near future. At the same time it could also be presumed that management of these technologies and their industrial promotion will condition further utilisation and development of nuclear power in a form acceptable to the public to a significant degree.

Experience with pyrochemical technologies in nuclear as well as in non-nuclear areas attained in the past indicates that fulfilment of this objective might be realisable by using these technologies.

REFERENCES

[1]   R.C. Robertson, ORNL 4541 (1971).

[2]   M. Bourgeois, B. Cochet-Muchy, *Bulletin d'Informations Scientifiques et Techniques*, No. 188 p. 41 (1971).

[3]   E. Yagi, S. Saito, M. Horiuchi, JAERI-M-6487 (1976).

[4]   P. Novy, NRI-9062 Ch (1989).

[5]   J. Uhlir, "Research and Development of Pyrochemical Technologies for ADTT Systems in NRI Řež plc", ADTTA'99, 3rd International Conference on Accelerator-Driven Transmutation Technologies and Applications, Prague, 7-11 June 1999.

ACTINIDE RECYCLING BY PYROPROCESS
FOR FUTURE NUCLEAR FUEL CYCLE SYSTEM

Tadashi Inoue
Central Research Institute of Electric Power Industry (CRIEPI)
2-11-1, Iwato-Kita, Komae-shi, Tokyo 201-8511, Japan
Tel.: 81-3-3480-2111, Fax: 81-3-3480-7956
E-mail: inouet@criepi.denken.or.jp

Abstract

Pyrometallurgical technology is one of the potential devices for the future nuclear fuel cycle. Not only economic advantage but also environmental safety and strong resistance for proliferation are required. So as to satisfy the requirements, actinide recycling applicable to LWR and FBR cycles by pyroprocess has been developed over a ten-year period at the CRIEPI. The main technology is electrorefining for U and Pu separation and reductive extraction for TRU separation, which can be applied on oxide fuels through reduction process as well as metal fuels. The application of this technology for separation of TRU in HLLW through chlorination could contribute to the improvement of public acceptance with regard to geologic disposal. The main achievements are summarised as follows:

- Elemental technologies such as electrorefining, reductive extraction, injection casting and salt waste treatment and solidification have been successfully developed with lots of experiments.

- Fuel dissolution into molten salt and uranium recovery on solid cathode for electrorefining has been demonstrated at an engineering scale facility in Argonne National Laboratory using spent fuels and at the CRIEPI through uranium tests.

- Single element tests using actinides showed Li reduction to be technically feasible; the subjects of technical feasibility on multi-element systems and on effective recycle of Li by electrolysis of Li_2O remain to be addressed.

- Concerning the treatment of HLLW for actinide separation, the conversion to chlorides through oxides has also been established through uranium tests.

- It is confirmed that more than 99% of TRU nuclides can be recovered from high-level liquid waste by TRU tests.

- Through these studies, the process flowsheets for reprocessing of metal and oxide fuels and for partitioning of TRU separation have been established.

The subjects to be emphasised for further development are classified into three categories: process development (demonstration), technology for engineering development and supplemental technology. In order to establish the technology, international collaboration must be substantial and effective.

Introduction

Since 1985, the CRIEPI has been developing:

- Metal fuel FBR cycle technology combined with pyrometallurgical reprocessing which has great potential with regard to environmental safety, nuclear non-proliferation and high economical efficiency.

- Reduction technology of U, Pu and MA (Np, Am, Cm) oxides in light water spent fuels to metallic form aiming at applying the pyrometallurgical technology (Li reduction process).

- TRU partitioning and transmutation technology used pyrometallurgy in order to increase public acceptance of high-level liquid waste management.

In the proposed advanced nuclear fuel cycle, recovered TRU nuclides will be mixed with U-Pu-Zr to make fresh fuels for FBR and will take a role of energy production. Consequently, trace amounts of TRU will be released into the environment. Integrating all of the technologies described above, the CRIEPI carries out research and development in the domain of "Actinide Recycling Technology by Pyroprocess" [1].

As concerns pyrometallurgical reprocessing for metal fuel, the CRIEPI took part in the integral fast reactor (IFR) project of the US DOE (ANL) from 1989 to 1995 [2]. The DOE stopped the IFR project in 1995 as a result of the political change banning the use of Pu. While the CRIEPI has continued to carry out R&D concerning pyrometallurgical technology integrated LWR and FBR fuel cycle, as actinide recycling by pyroprocess is called, in-house activity has been aided by domestic and overseas institutes. In the integrated system, the CRIEPI carried out the research and development of TRU partitioning and transmutation based on the original concept [3], which is nominated to the OMEGA project (long-term research and development programme for the partitioning and transmutation of long-lived nuclides) in Japan. The programmes, including those of JAERI, JNC and CRIEPI, were checked and reviewed by the Atomic Energy Commission of Japan in 1999.

Characteristics of the system with pyroprocess

Characteristics on nuclear fuel recycle scenario

Pyroprocess technology:

- Can be applied on any type and any composition of nuclear fuels, for instance, not only on metal fuel but also on oxide fuel and nitride fuel.

- Is applied independent of burn-up of spent fuel and even of the short cooling fuels.

- Possesses a high resistance with regard to nuclear proliferation.

- Has a high potential with regard to environmental safety, because the release of actinides is limited to trace amounts.

The pyroprocess facility is well fitted to collocate with reactors, which would greatly reduce the transport of nuclear materials, once the integrated system is realised.

Characteristics of technology

The process for reprocessing and recovering minor actinides is expected to be simple and will consist of a compacted facility and equipment. A small amount of waste will be constantly generated under normal operation, because neither organic solvent nor nitric acid solution is used. A large margin for criticality will be relatively expected, because of the lack of an aqueous phase. The proposed TRU recovery process can be applicable not only to high-level liquid waste but also to the slurry, undissolved residue and solvent washing reagent in wastes coming from aqueous reprocessing. On the other hand, it should be noted that compared with PUREX process, less experience has been accumulated with the pyroprocess. A large amount of effort will be required for commercial realisation.

Outlook of pyroprocess technology

U and Pu electrorefining process

The metal fuel (U-Pu-Zr) is dissolved into molten salt electrolyte and uranium, plutonium and some part of minor actinides through electrorefining. The advantage of the electrorefining is that the process flow becomes very simple because almost all of the actinide elements will be collected together.

Reduction process of oxide fuel

The spent oxide fuels discharged from light water reactors are reduced to metals in a molten LiCl system using Li metal as a reductant. Adding to this process, pyrometallurgical reprocessing will be applied to oxide fuels.

Chlorination process of high-level waste

In order to apply the pyroprocess for separating TRUs from high-level liquid waste resulting from PUREX reprocessing, the liquid waste of nitric solution must be converted to chloride through oxide formation by denitration. Chloride elements will undergo the TRU reductive extraction process.

Reductive extraction process for recovery of TRUs

TRUs in salt phase are extracted into liquid metal by using Li metal as a reductant. Separation efficiency between TRUs and FPs depends on the difference of formation free energy of chlorides.

Waste treatment process for chloride salt

Chloride salt waste containing highly radioactive FP after recovering U, Pu and MA is converted into stable form, such as synthetic rock or borosilicate glass, which can be disposed of in geologic formations. The generation of secondary radioactive waste will be minimised by recycling eutectic salt and reductant in the process.

Injection casting of metallic fuels

Fuel slag is cast into glass moulds immediately by pressuring the molten metals after adjusting the component of U, Pu, MA and Zr. Lots of fuels can be fabricated at once in a compacted facility.

Supplemental technology

In order to introduce pyroprocess technology at a practical level, remote control technology, molten salt and liquid metal transportation technology and nuclear material safeguard technology must be developed.

Review of achievements

U and Pu electrorefining process

Development of electrorefining technology at ANL has advanced to laboratory scale testing and engineering scale demonstration tests [4]. Due to the policy change excluding the use of Pu, a demonstration of the complete process of pyroreprocessing has not been realised up to now. The technologies, which have been demonstrated at engineering scale, include the spent fuel dissolution process, U collection process by electrorefining on the solid cathode and U separation process from adhered salt.

At the CRIEPI, the optimisation of operating condition with regard to electrorefining is investigated. Uranium tests on the scale of 1 kg scale have been conducted [5,6]. Progress in the domain of electrorefining with solid cathode and liquid cathode has been made as a result of a joint study with JAERI using Pu and Np [7].

Reduction process of oxide fuel

ANL and CRIEPI, independently, confirmed that more than 99% of UO_2 was converted into metal form [8]. The experiments used single elements of PuO_2, Am_2O_3 or NpO_2 and used MOX pellet. The testing revealed the feasibility of conversion to metals, which was jointly conducted by CRIEPI and AEAT [9]. Furthermore, a demonstration test using simulated spent fuel will begin this year.

Chlorination process of high-level liquid waste

A demonstration test using simulated high-level waste indicates that almost all rare earths, noble metals and actinides are denitrated, remaining alkali nitrates at 500°C [10]. In this step, most of the alkali elements of FP with high heat emission can be separated by water rinsing after denitration. The oxides formed are converted to chlorides by pouring chlorine gas with carbon reductant over 700°C in a LiCl-KCl bath [10]. Some of the chlorides evaporated are captured in a LiCl-KCl bath kept at a lower temperature.

Reductive extraction process for recovery of TRUs

Prior to the separation experiments, free energies of formation of chlorides were obtained from the standard potentials measured electrochemically for rare earths and actinides [11]. The distribution coefficients of each element were also measured in systems of LiCl-KCl/Cd and LiCl-KCl/Bi [12].

Experiments for separation of TRUs were carried out and the recovery yields of all actinides were found to be over 99% [13].

The optimised TRU separation process flowsheet is established when more than 99% TRU is separated, maintaining a large separation from rare earths, and solvent salt, reductant and Cl_2 gas can be recycled in the process [14].

Waste treatment process for pyroprocess

Two kinds of waste treatment process have been developed.

Firstly, heated $NaAlO_2$ and SiO_2 with waste salts at high temperature, the sodalite structure is synthesised, in which waste chlorides are immobilised in an artificial mineral matrix [15].

The second process is to form borosilicate glass contained waste. Salt waste is converted to oxides through the step of electroreduction in liquid lead and vitrified with B_2O_3 and SiO_2 at air atmosphere. Most of all solvent salt and solvent metal after electrorefining can be recycled, and Cl_2 gas generated in this process will be used for the chlorination process [16].

Injection casting of metal fuels

ANL has already developed the technology to fabricate fuel at an engineering scale. In CRIEPI, the engineering-scale injection casting test has begun.

Summary

Actinide recycling technology by pyroprocess has been under development at CRIEPI since 1985, with the aim of promoting the next generation nuclear fuel cycle system with high potentials of environmental safety, non-proliferation and economic advantage. Through studies, elemental technologies, such as electrorefining, reductive extraction, injection casting, and salt waste treatment and solidification, have been successfully developed. The fuel dissolution into molten salt and uranium recovery on solid cathode for electrorefining have been demonstrated with spent fuels at the engineering scale facility in ANL, and at CRIEPI through uranium tests. Single element tests using actinides showed the Li reduction to be technically feasible. The subjects of applicability on multi-element systems and effective recycle of Li by electrolysis of Li_2O remain to be addressed. Concerning the treatment of HLLW for actinide separation, the conversion to chlorides through oxides has also been established through uranium tests. Through testing, it has been confirmed that more than 99% of TRU nuclides can be recovered from high-level liquid waste. These studies have helped establish the process flowsheets for reprocessing of metal and oxide fuels and for partitioning of TRU separation.

However, the pyroprocess has been much less experienced with as compared to the aqueous process. Further efforts are required from the engineering points of view, for example removal or dissolution of dross accumulated in equipments, handling of the cathode product and separation from salt, collection of dropped material from cathode product, transportation of the melt and remote operation technology. In addition, an accounting measure of fissile material needs to be developed, according to batch-wise operation. Material selection for high temperature operation is not the least issue.

REFERENCES

[1]   T. Inoue and H. Tanaka, "Recycling of Actinides Produced in LWR and FBR Fuel Cycles by Applying Pyrometallurgical Process", Proc. on Future Nuclear Systems (GLOBAL'97), pp. 646-652, 5-10 Oct. 1997, Yokohama, Japan (1997).

[2]   Y.I. Chang, "The Integral Fast Reactor", *Nucl. Technol.*, 88, 129 (1989).

[3]   T. Inoue, M. Sakata, H. Miyashiro, T. Matsumura, A. Sasahara, N. Yoshiki, "Development of Partitioning and Transmutation Technology for Long-Lived Nuclides", *Nucl. Technol.*, 69, 206 (1991).

[4]   T.C. Totemeier and R.D. Mariani, "Morphologies of Uranium and Uranium-Zirconium Electrodeposits", *J. Nucl. Mater.* 250, 131 (1997).

[5]   T. Koyama, M. Iizuka, Y. Shoji, R. Fujita, H. Tanaka, T. Kobayashi and M. Tokiwai, "An Experimental Study of Molten Salt Electrorefining of Uranium Using Solid Iron Cathode and Liquid Cadmium Cathode for Development of Pyrometallurgical Reprocessing", *J. Nucl. Sci. Technol.*, 34, 384 (1997).

[6]   T. Koyama, M. Iizuka, N. Kondo, R. Fujita and H. Tanaka, "Electrodeposition of Uranium in Stirred Liquid Cadmium Cathode", *J. Nucl. Mater.*, 247, 227 (1997).

[7]   Y. Sakamura, O. Shirai, T. Iwai and Y. Suzuki, "Thermodynamics of Neptunium in LiCl-KCl Eutectic/Liquid Bismuth Systems", *J. Electrochem. Soc.*, in press.

[8]   E.J. Karell, K.V. Gourishankar, L.S. Chow and R.E. Everhart, "Electrometallurgical Treatment of Oxide Spent Fuels", Int. Conf. on Future Nuclear Systems (GLOBAL'99), Jackson Hole, Wyoming, 29 Aug.-3 Sept. (1999).

[9]   T. Usami, M. Kurata, T. Inoue, J. Jenkins, H. Sims, S. Beetham and D. Browm, "Pyrometallurgical Reduction of Unirradiated TRU Oxides by Lithium in a Lithium Chloride Medium", OECD Nuclear Energy Agency, Nuclear Science Committee, Workshop on Pyrochemical Separations, 14-15 March 2000, Avignon, France (2000).

[10]   M. Kurata, T. Kato, K. Kinoshita and T. Inoue, "Conversion of High-Level Waste to Chloride for Pyrometallurgical Partitioning of Minor Actinides", Proc. of 7th Int. Conf. on Radioactive Waste Management and Environmental Remediation, ICEM'99, 26-30 Sept. 1999, Nagoya, (1999).

[11]   Y. Sakamura, T. Hijikata, K. Kinoshita, T. Inoue, T.S. Storvick, C.L. Krueger, J.J. Roy, D.L. Grimmett, S.P. Fusselman and R.L. Gay, "Measurement of Standard Potentials of Actinides (U,Np,Pu,Am) in LiCl-KCl Eutectic Salt and Separation of Actinides from Rare Earths by Electrorefining", *J. Alloy. Compound*, 271-273, 592 (1998).

[12] M. Kurata, Y. Sakamura and T. Matsui, "Thermodynamic Quantities of Actinides and Rare Earth Elements in Liquid Bismuth and Cadmium", *J. Alloy. Compound*, 234, 83 (1996).

[13] K. Kinoshita, T. Inoue, S.P. Fusselman, D.L. Grimmett, J.J. Roy, R.L. Gay, C.L. Krueger, C.R. Nabelek and T.S. Storvick, "Separation of Uranium and Transuranic Elements from Rare Earth Elements by Means of Multistage Extraction in LiCl-KCl/Bi System", *J. Nucl. Sci. Technol.*, 36, 189 (1999).

[14] K. Kinoshita, M. Kurata and T. Inoue, "Estimation of Materials Balance in Pyrometallurgical Partitioning Process of Transuranic Elements from High-Level Liquid Waste", *J. Nucl. Sci. Technol.*, 37, 75 (2000).

[15] T. Koyama, C. Seto, T. Yoshida, F. Kawamura and H. Tanaka, "Immobilization of Halide Salt Waste from Pyrochemical Reprocessing by Forming Natural Occurring Mineral; SODALITE", Int. Conf. on Evaluation of Emerging Nuclear Fuel Cycle Systems (GLOBAL'95), Vol. 2, 1744, 11-14 Sept. 1995, Versailles (1995).

[16] Y. Sakamura, T. Inoue, T. Shimizu and K. Kobayashi, "Development of Pyrometallurgical Partitioning Technology for TRU in High-Level Radioactive Wastes – Vitrification Process for Salt Wastes", Int. Conf. on Future Nuclear Fuel Systems (GLOBAL'97), Vol. 2, 1222, 5-10 Oct. 1997, Yokohama (1997).

EVALUATION OF PYROCHEMICAL REPROCESSING TECHNOLOGIES FOR PRACTICAL USE

Shinichi Kitawaki, Kazumsa Kosugi, Mineo Fukushima, Munetaka Myochin
Advanced Fuel Recycle Technology Division
Japan Nuclear Cycle Development Institute
4-33, Muramatsu, Tokai-mura, Naka-gun, Ibaraki,319-114, Japan
Tel.: 81-029-282-1111, Fax: 81-029-282-0864

Abstract

The investigation and evaluation of pyroprocess technology were performed to clarify the problems for future nuclear fuel cycle technology. The oxide electrowinning method developed at RIAR and the metal electrorefining method developed at ANL have been investigated. It is evident that both methods are almost at engineering scale level. Many difficulties exist, however, with regard to the fundamental/industrialisation technology required to construct such facilities in Japan.

JNC is currently making an effort to co-operate with the electric utilities to resolve these problems for the future nuclear fuel cycle.

Introduction

Due to its simplicity, it is expected that pyrochemical reprocessing will be a more economic process as compared to the aqueous method. The Japan Nuclear Cycle Development Institute (JNC) is currently evaluating the metal fuel electrorefining method (developed in United States ANL) and the oxide fuel electrowinning method (developed in Russia RIAR) in co-operation with the electric utility on the assumption of the application of the pyrochemical reprocessing to the oxide fuel cycle. This paper discusses an investigation result based on domestic and overseas information, the present evaluation and the development problem in the future.

Background

In recent years, from the perspective of the economy of the nuclear fuel cycle and the reduction of environmental impact, attention has been paid to confining FP at the cycle and to burning FP in FBR. This technology is advantageous in that U/Pu need not be separated as with a conventional aqueous reprocessing method for high decontamination factor (DF). A low DF is a factor which improves the economy of the process, as it is necessary to operate the fuel fabrication process remotely.

In 1983, the IFR programme was proposed to achieve dry recycling through pyrometallurgy at the metal fuel fast reactor EBR-II and FCF in United States at ANL. Afterwards, the IFR plan was cancelled due to a political judgement concerning Pu in 1994. The engineering scale equipment is currently driven under the DOE Spent Fuel Treatment Programme. The feature of this process is to deposit the metal by electrorefining, and to produce the fuel element using the injection casting method technology [1].

The development of an engineering scale oxide fuel fast reactor and oxide recycling technology has been continuing at Russia's RIAR since 1963. The main principal of this process is to deposit the oxide grain by electrowinning, and to produce the fuel element from the vibration compaction method.

Each technology adopts molten salt electrolysis for the separation step and simple fuel fabrication technology which is adapted to the reactor. Both processes have economically achieved the fuel cycle where FP is confined [2,3].

Investigation

The present state and the problem of pyrochemical process were investigated from public information obtained now. The investigated items were classified into three categories: recycling technology which includes reprocessing and fuel fabrication, system technology to make industrial scale and waste disposal technology.

Recycling technology

Metal fuel electrorefining [1]

Metal fuel electrorefining has already been examined in engineering scale, and results from ANL demonstrate the feasibility of the metal fuel cycle. Metal fuel electrorefining uses solid cathode to deposit uranium. The engineering scale test showed a 10 kg uranium deposit. Also, a new electrorefiner is being developed to improve the processing speed. The ability to recover plutonium by liquid Cd cathode is confirmed at laboratory scale examination. Moreover, the cathode deposition is processed by the distillation, and an excellent separation of metal and salt is confirmed. The nuclear material is processed to the fuel element through injection casting.

Oxide fuel electrowinning

In Russia, manufacturing results of the exchange fuel for a fast reactor (mainly BOR-60) are available. This method was developed for fuel fabrication, and reaches up to engineering scale. The problem is the amount of time required when the oxide is dissolved in the molten salt. In the oxide fuel electrowinning of RIAR, only Pu is recycled because UO_2 is deposited with the main FP, and U is abandoned. The precipitation of Pu by pouring the mixed gas of oxygen and chlorine into the salt after separation of the uranium is adopted. Pu of about 500 g is collected by the examination in which spent fuel of the burn-up 24% is used [4]. Precipitation and deposit are washed in water for fuel fabrication and the obtained grain is crushed, and classified. Then grain is packed cladding tube being vibrated. The uranium metal is used as a getter material for the preparation of O/M rate [6].

System technology

System technology is composed of safety, process control, operation, etc., and it is basically common technology to both methods. In the current investigation, the majority of system technologies were solved from performance of the metal fuel electrorefining and the oxide fuel electrowinning has already processed tonnes of fuel in total. Then, the problems to apply the method in Japan were enumerated.

Multi-line of equipment

Though both metal fuel electrorefining and oxide fuel electrowinning have results, when the plant scale is large, the multi-line of the electrorefiner is needed, and the operation management of the plant becomes complex. Moreover, the handling technology of solids such as cathodes and crucibles, and fine particle or grain such as fuel element is used frequently. There is a possibility that reliability and maintainability become problems, though the experience of the aqueous process can be applied [6].

Safeguards technology

The material accounting is adopted as a main means of safeguards in Japan. Therefore, the input accountancy after the fuel dissolution is performed in process. However, since the fuel dissolution and the cathode deposition are performed at the same time in metal fuel electrorefining, the same concept applied for aqueous process cannot be used here. Due to a shortage of information, the current state of the oxide method is not fully understood. Moreover, it is difficult to obtain a uniform sample from molten salt in pyroprocess.

Safety data

To acquire a government licence as concerns safety design and evaluation of a pyroprocess, physical/chemistry data is necessary to explain the behaviour of abnormal condition, in addition to the behaviour of normal operation. The data is insufficient compared with aqueous process which is reaching a commercial base now.

Atmosphere control

For metal fuel electrorefining, it is necessary to keep a high-grade argon atmosphere in the cell. For oxide fuel electrowinning, it is necessary to keep a dried air atmosphere in the cell. Atmosphere control is expected to increase the difficulty as a result of an increase in the number of mechanical openings and expansion of the facilities' scale (laboratory scale, engineering scale and commercial scale).

Material corrosion

A definite difference of the pyroprocess is its high reaction temperature as compared to an aqueous process. The temperature reaches about 1 300°C in the cathode process in the metal fuel electrorefining. A ZrO_2 coating graphite crucible is used as a corrosion resistant material. As for oxide fuel electrowinning, though pyrographite has been selected as a material which possesses corrosion resistance in the chlorine gas environment, its lifetime is assumed to be only about 1 000 hours. Additionally, excluding a main process, the lack of a corrosion resistant material in the off gas processor has already been pointed out for both methods.

Waste disposal technology

The waste salt processing technique of the sodalite solidification is proposed in the United States, and the manufacturing technology of the vitrified waste in the engineering scale development is being performed now. In Russia the waste salt is stored as is. As concerns phosphate precipitation, which contains a large amount of FP from electrowinning, the solidification technology of the phosphate glass has been developed [7]. Moreover, since the chlorine gas is used voluminously in the oxide fuel electrowinning, recycling and collecting technology of chlorine gas is necessary.

Evaluation viewpoint

JNC studies the candidate technology for the FBR cycle (which can rival the light water reactor, in regard to economy) in the Feasibility and Strategy Study Programme. To clarify technical approval and the development problem, technology is evaluated from viewpoints of economy, resource effective use, reduction of environment impact, and non-proliferation of nuclear weapons on the assumption of safety. Pyrochemical reprocessing technology is evaluated in the following categories.

Fuel type

Metal and nitride fuels have the possibility to have a greater economical and technical potential than the oxide fuel. However, most nuclear reactors, excluding research reactors, adopt oxide fuel. The process for commercial reprocessing is aqueous method for oxide fuel. Therefore, to apply to a present nuclear fuel cycle system including the waste disposal, oxide fuel is targeted.

Effective use of resources

The effective use of not only Pu resource but also U is needed in the FBR cycle. Therefore, it is preferable to raise recovery of plutonium and uranium as much as possible on industrialising. It is assumed 99%-99.9% [8] as standard recovery by which the feature of a pyroprocess is hardly lost.

Reduction of environment impact

In reduction of environment impact, confining a lot of radionuclides in the system conflicts with the economy. Moreover, it is necessary for selecting recycle nuclide to consider amount of the nuclide, long-term radioactive, and long-term toxicity, etc. In this paper, however, the recovery of MA is not considered so that separating a specific nuclide may lose the feature of simplicity of the pyroprocess.

Non-proliferation of nuclear weapons

There is presently no international standard concerning the non-proliferation of nuclear weapons. In this paper, the standard that pure plutonium cannot exist alone in each process of the fuel cycle is applied.

Safeguards

The material accounting is performed as a main means for safeguards. As for the strict evaluation of safeguards, after the facilities are designed and divided into two or more material balance areas, the evaluation can be performed. For aqueous reprocessing, input accountancy can be performed by measuring homogeneous solution in which the whole quantity of nuclear material has been dissolved. However, input accountancy of the electrolytic method is difficult, because nuclear material deposits to the electrode while being dissolved to the molten salt. Since facilities have not yet been designed, the applicability of input accountancy is one of evaluation viewpoint.

Safety data

If pyroprocess facilities are constructed in Japan, the procedure based on present law (applied to reprocessing facilities and fuel fabrication facility in Japan) and license system will have to be used. To acquire a government licence by safety design and assessment of pyroprocess, physical/chemistry data is necessary to explain the behaviour of abnormal condition, in addition to the behaviour of normal operation. The data is insufficient compared to the aqueous process which is already reaching a commercial base.

Waste disposal

The feature of waste from pyroprocess is chloride salt. As described in the section on target fuel, not only in Japan but also in many countries, borosilicate vitrification of high-level radioactive liquid waste is examined. In disposal of chloride waste, it is necessary that the development be done over again by the method same as a present vitrified waste or the chloride waste is treated to dispose by a present method.

Evaluation result

The molten salt electrolysis examinations used spent fuel are performed, and a lot of results of the fuel fabrication technologies have been obtained. However, it is necessary to overcome the following basic technological/industrialisation problems (listed in Tables 1 and 2) for applying the pyrochemical reprocessing technology in Japan. Figures 1 and 2 show the flowsheet of oxide fuel electrowinning (developed in RIAR) and metal fuel electrorefining (developed in ANL) applied in Japan respectively.

Recycling technology

As for oxide fuel electrowinning, the recycling technology of uranium which is hardly recycled in Russia, and separation technology of collected UO_2 and FP is important in Japan. If the U/Pu codeposition technology developed as a method of the replacement to precipitation is approved to the recovery of Pu, fuel fabrication becomes easy and the process can be speeded up.

It is necessary to develop "reduction" and "oxidation" processes for applying metal fuel electrorefining to the oxide fuel. Moreover, it is necessary to develop the recycling technology of Li_2O generated from Li reduction and removal technology of FP in the salt bath for the decrease of waste. Though the oxidation process can be easily achieved because of simple principle, it is necessary to select the oxidation method in consideration of the correspondence with the fuel fabrication.

Waste technology

The problem of waste is common to both methods.

- *Approval of salt recycling technology*. A large amount of FP in the spent fuel is involved to the molten salt used for electrolysis. Therefore, the FP recovery technique is necessary to recycle salt.

- *Overcoming material corrosion*. In pyroprocess, the problem of the material corrosion relates directly to that of waste. In the oxide fuel electrowinning, the life prolongment of the pyrographite crucible or finding the substitution material is hoped for. In the metal fuel electrorefining, equipment material for Li_2O reduction process is not found.

- *Approval of salt processing technology*. Salt is basically reused but it finally becomes waste. Similar solidification technology and performance evaluation similar to the vitrified waste of aqueous method are necessary.

System technology

It is difficult for pyroprocess to measure with a homogeneous solution in which the whole quantity of nuclear material is dissolved. As a method of the input accountancy, it is necessary to analyse and sample the powder which is produced with dismantling/decladding of the fuel and mixed sufficiently to be homogeneous. However, it is necessary to discuss the safeguards approach with other countries in the future. Moreover, the method of confirming the behaviour of Pu in the process is necessary from the viewpoint of production and criticality control. Therefore, it is necessary to understand not only the behaviour of plutonium but also the behaviour of the FP nuclide.

Industrialisation problem

The following issues need to be resolved with regard to the industrialisation of the pyrochemical process.

Improvement of recovery

Pyrochemical process has been developed mainly with molten salt electrolysis process, and has been hardly studied in head-end and decladding process. "Voloxidation" by expansion of UO_2 volume and "mechanical decladding" by mechanical vibration or pressure are considered. It is necessary to examine the method of recovery of crushed powder fuel and adhering to the cladding tube.

Collection of safety data

The following data should be examined: basic behaviour and recovery of FP, thermal behaviour of the equipment, assumption of the incident peculiar to pyrochemical process, etc.

Analysis, measurement and control technology for process management

One of the disadvantages of pyrochemical reprocessing is the length of time required. In addition, because pyrochemical process is basically composed of batch processing, an increase in the analysis point is expected. It is necessary to apply the means of prompt analysis and to develop the technology of *in situ* spectrum analysis. Moreover, measurement of temperature, level, pressure and density of molten salt and metal is needed under the condition where a large amount of FP exists. The accumulation of these technologies is needed.

Material handling

A lot of handling solid material by remote control will be applied in pyrochemical process because of batch processing. Therefore, it is expected that the installation cost increases and the equipment reliability decreases by making the multi-line and remote control of pyroprocess equipment.

Atmosphere control

It will be a problem to design equipment which is sufficiently airtight for the atmosphere control, and it will also be difficult to finance the extra cost that such a structure will impose.

As a means of addressing the above-mentioned problem, JNC will develop the component and system technology of pyrochemical process in co-operation with electric utilities. There are many problems which still need to be solved, and it is necessary to do a large effort for practical use.

REFERENCES

[1] "Metal Fuel Cycle Technology – Present Status and Future Perspective", Atomic Energy Society of Japan 1995.10.

[2] O.V. Skiba, Yu.P. Savochkin, A.V. Bychkov, P.T. Porodnov, L.G. Babikov and S.K. Vavilov, "Technology of Pyrochemical Reprocessing and Production of Nuclear Fuel", GLOBAL'93.

[3] V.B. Ivanov, O.V. Skiba, A.A. Mayershin, A.V. Bychkov, L.S. Demidova, P.T. Porodnov, "Experimental, Economical and Ecological Substantiation of Fuel Cycle Based on Pyroelectrochemical Reprocessing and Vibropac Technology", GLOBAL'97.

[4] A.V. Bychkov, S.K. Vavilov, O.V. Skiba, P.T. Porodnov, A.K. Pravdin, G.P. Popkov, K. Suzuki, Y. Shoji, T. Kobayashi, "Pyroelectrochemical Reprocessing of Irradiated FBR MOX Fuel. III. Experiment on High Burn-Up Fuel of the BOR-60 Reactor".

[5] V.B. Ivanov, A.A. Mayershin, O.V. Skiba, V.A. Tzykanov, G.I. Gadgiev, P.T. Porodnov, A.V. Bychkov, V.A. Kisly, "The Utilisation of Plutonium in Nuclear Reactors on the Basis of Technologies Developed in SSC RIAR".

[6] W.R. Bond, G. Jansen, Jr., L.K. Mudge, "Demonstration of the Salt Cycle Process in a High-Level Radioactive Facility", AEC Research and Development Report, BNWL-355.

[7] A.N. Lukinykh, Yu.G. Lavrinovich, A.V. Bychkov, O.V. Skiba, S.K. Vavilov, Y.A. Erin and S.V. Tomilin, "High-Level Waste Composition, Properties and Preparation for Long-Term Storage after Pyroelectrochemical Reprocessing of Irradiated Oxide Uranium-Plutonium Fuel", RECOD'98.

[8] K.M. Harmon and G. Jansen, Jr., "The Salt Cycle Process", *Progress in Nuclear Energy Series III*, 4 (1970) 429.

Figure 1. Pyrochemical flowsheet for oxide fuel recycle based RIAR

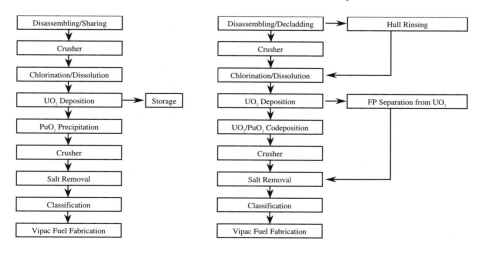

RIAR   Modified RIAR

Figure 2. Pyrochemical flowsheet for oxide fuel recycle based ANL

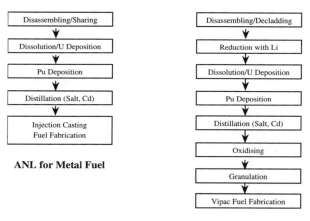

ANL for Metal Fuel   ANL for Oxide Fuel

Table 1. Oxide fuel electrowinning

		R&D items
Feasibility problem	Recycling process technology (*reprocessing/fuel fabrication*)	Decontamination process of FP from uranium deposit Technique of decladding UO_2, PuO_2 co-deposition
	Waste technology	Development of waste treatment technology
	System technology	Material accountancy system
Industrialisation problem	Recycling process technology (*reprocessing/fuel fabrication*)	Improvement of recovery Material corrosion (*ex. pyrographite crucible*) Technique of pin examination (*ex. Pu distribution, QA etc.*) Data for mass flow estimation
	Waste technology	Recycling process of used salt Estimation of amount of wastes
	System technology	Safety evaluation for Japanese regulation Approval of process management system (*accountancy, criticality, quality of products etc.*) Design of instrument (*off gas system, material handling, chlorine recycling system etc.*)

Table 2. Metal fuel electrorefining

		R&D items
Feasibility problem	Recycling process technology (*reprocessing/fuel fabrication*)	Lithium reduction technique of MOX Restrain of eutectic reaction between fuel and cladding tube
	Waste technology	Development of waste treatment technology
	System technology	Material accountancy system
Industrialisation problem	Recycling process technology (*reprocessing/fuel fabrication*)	Cadmium cathode for industrial-scale Performance estimation of anode-cathode module Material corrosion (*ex. crucible*)
	Waste technology	Salt recycling technology Waste management technology Estimation of amount of wastes
	System technology	Design studies in commercial scale Safety evaluation for Japanese regulation (*accountancy, criticality, quality of products etc.*)

SESSION III

Part A: Basic Data (Thermodynamics, Reaction Mechanisms, Reaction Kinetics, Molecular Modelling, etc.)

Chairs: L. Koch, T. Ogawa

MOLTEN SALT DATA FOR PYROCHEMISTRY: NEW RESEARCH ADVANCES AND DEVELOPMENT OF INFORMATION SYSTEM

M. Gaune-Escard
[a] IUSTI, CNRS UMR 6595, Technopôle de Chateau Gombert
5 rue Enrico Fermi,13453 Marseille Cedex 13, France

A.K. Adya
[b] Division of Molecular and Life Sciences, School of Science and Engineering
University of Abertay Dundee, Bell Street, Dundee DD1 1HG, Scotland

L. Rycerz [a,c]
[c] Institute of Inorganic Chemistry and Metallurgy of Rare Elements
Technical University Wybrzeze Wyspianskiego 27, 50-370 Wroclaw, Poland

Abstract

Pyrochemical reprocessing of nuclear fuels simultaneously involves materials, energy production and environmental aspects. However, the data relative to molten lanthanide and actinide salts (and compounds), specific to this field, are not easily available. Because of characteristic physicochemical properties, which do not promote experimental investigations, data are scarce. Very often, most of the data is buried in grey literature. When available, they reveal to arise from estimations. Co-ordination and standardisation of existing data is therefore essential, as research efforts should be intensified together with development of numerical prediction tools. This paper aims, firstly, to discuss the new experimental data obtained through an intensive international research co-operation and, secondly, to present ongoing efforts for the development of a molten salt database.

Introduction

Systematic investigations were recently conducted on molten trivalent halides, which are important from both the technological and scientific point of view. Their thermodynamic, structural and transport properties are determined by several complementary experimental methods, while modelling and numerical simulations give more insight both about the microscopic and macroscopic behaviour [1-8]. These melts constitute a challenging class of systems with regard to modelling the interatomic interactions. The objective is to develop a generalised ionic model which is easily transferable between closely related, chemically similar materials, mainly through a change in the cation radius.

We report in the following as examples some recent results obtained on:

- The thermodynamic properties of pure europium trihalide $EuCl_3$, which complement previous investigations performed on europium dihalide $EuCl_2$ [9].

- The structure of molten UCl_3 by neutron diffraction.

- The phase equilibrium calculation of UCl_3-NaCl melt.

Some experimental aspects, of importance for the preparation of high purity lanthanide samples, are also emphasised beforehand.

Thermodynamics of EuCl₃

The experimental methods used for thermodynamic of characterisation and samples preparation are described in the following. Enthalpy and heat capacity results are detailed, as an example, for $EuCl_3$.

Experimental

The enthalpies of phase transitions and heat capacities were measured with a SETARAM DSC 121 differential scanning calorimeter. The apparatus and the measuring procedure were previously described in detail [10].

The so-called "step method" used for C_p measurements has already been described [5,10,11]. With this method, small heating steps are followed by isothermal delays, when thermal equilibrium of the sample is achieved. Two correlated experiments should be carried out with this method to determine the heat capacity of the sample. The first one should be performed with two empty cells (containers) having identical mass, and the second with one of the same cells with the sample. The heat flux as a function of time and of temperature is recorded for both runs. The difference of heat flux in each run is proportional to the amount of heat (Q_i) necessary to increase the temperature of the sample by •T_i. Therefore, the heat capacity of the sample (C_p) is equal to:

$$C_p = \frac{Q_i \cdot M_s}{\Delta T_i \cdot m_s}$$

where m_s is the mass of the sample, and M_s is the molar mass of the sample.

Sample preparation

Because of the high reactivity of lanthanide chlorides, preparation of pure and water-free samples is essential. Much attention should be paid in order to avoid melt contamination by oxichloride. We will detail in the oral presentation the procedures followed for the preparation of those "stable" trivalent halides and those less stable "SmCl$_3$ and EuCl$_3$, which undergo partial thermal decomposition upon melting.

The chemical analysis of the synthesised lanthanide chlorides was performed by titration methods for chloride (mercurimetric) and lanthanide (complexometric).

Enthalpy of phase transition

The melting temperature of EuCl$_3$ determined from our experiments is in a good agreement with the data of Kulagin and Laptev [12] and Moriatry [13] (894, 897 and 896 K, respectively). Other literature data [14] give the value of 901 K as the melting temperature of EuCl$_3$, but it was obtained by an extrapolation of experimental results of partially dissociated compound. Our determination of the molar enthalpy of fusion •$_{fus}H_m$ = 50.3 k J mol^{-1} is about 10% less than literature data [12].

However, the thermograms observed during successive heating-cooling cycles showed that upon primary heating, one single endothermic effect takes place at T$_{ons}$ = 894 K (melting of EuCl$_3$) while an additional effect appears at T$_{ons}$ = 796 K upon cooling (Figure 2). Upon further heating, two effects were observed at 804 K and 887 K, respectively (Figure 3). Chemical analysis of EuCl$_3$ samples after primary heating showed reduced chlorine content with respect to the stoichiometric composition (EuCl$_{2.97}$ instead of EuCl$_3$). Visual observations also confirmed chlorine release above molten samples.

This thermal decomposition of EuCl$_3$ is likely to occur at temperatures very close to melting. The single peak observed upon primary heating does correspond to real melting of EuCl$_3$.

When EuCl$_3$ decomposition takes place, EuCl$_2$ is formed and the sample under investigation turns into a EuCl$_3$-EuCl$_2$ mixture. On subsequent heating thermogram two thermal effects appear. They can be assessed to the fusion of (i) the eutectic fraction at T = 804 K and (ii) the pure EuCl$_3$ component at T = 887 K in the EuCl$_3$-EuCl$_2$ system.

The latter temperature is not in contradiction with the value T = 894 K observed upon primary heating, as, because of decomposition, there is some uncertainty in the temperature extrapolation from experimental thermograms. The eutectic temperature is in good agreement with that reported earlier by Laptev, *et al.* [15]. The same authors also concluded upon the thermal decomposition of EuCl$_3$ upon melting [12].

Heat capacity

The only experimental heat capacity data available in literature were those obtained at low temperature by adiabatic calorimetry by Sommers and Westrum [16]. At high temperature, no direct C$_p$ values were reported. However, C$_p$ data have been tabulated over an extended temperature range in several thermochemical tables [17-20] but they correspond to estimations from the previous low temperature heat capacities.

The experimental results of heat capacity measurements for $EuCl_3$ are shown in Figure 4 and compared with literature data [16] or estimating [17]. Our results for solid $EuCl_3$ are generally lower than those reported by Sommers and Westrum [16]. This difference is about 8% at 300 K and 11% for 790 K. For liquid $EuCl_3$, we found $C_p = 155.96$ J mol^{-1}K^{-1}. This value is about 10% higher than that estimated by Pankratz [17]. These original C_p determinations support our previous conclusion concerning europium trichloride decomposition at higher temperatures. Indeed the sudden temperature increase of the C_p at 802 K up to 160 J mol^{-1}K^{-1} (arrow 3 in Figure 4) corresponds exactly to the temperature of the additional thermal effect (804 K) observed on the DTA curves during a secondary heating of $EuCl_3$ (Figure 4).

Results and discussion

In 1971 Dworkin and Bredig [21] made a correlation between the crystal structure of lanthanide chlorides and their entropy of melting. They found that lanthanide chlorides with $Y(OH)_3$-type structure ($LaCl_3$, $CeCl_3$, $PrCl_3$, $NdCl_3$ and $GdCl_3$) have an entropy of melting $\bullet_{fus}S_m$ of about 50 ± 4 J mol^{-1}K^{-1}, whereas for the lanthanide chlorides with the $AlCl_3$-type structure ($DyCl_3$, $ErCl_3$) this entropy, or the sum of the entropies of transition and of fusion ($\bullet_{trs}S_m + \bullet_{fus}S_m$), is significantly lower and is equal to 31 ± 4 J mol^{-1}K^{-1}.

Tosi, *et al.* [22] generalised these observations and proposed that the melting mechanism of trivalent metal chlorides could be classified into three main types in correlation with the character of the chemical bond. The $AlCl_3$-type structure is layered and can almost be viewed as a cubic close packing of Cl ions inside which the metal ions occupy suitable octahedral sites. The UCl_3-type structure (also known as the $Y(OH)_3$-type structure) is described as hexagonal, with each U surrounded by six Cl on the corner of a trigonal prism and further co-ordinated by the three coplanar Cl at somewhat larger distance. The $PuBr_3$-type structure appears to be of a transitional type between the $AlCl_3$- and UCl_3-type structures.

The melting from the UCl_3-type or $PuBr_3$-type structure ($TbCl_3$) involves appreciably higher entropy than from the $AlCl_3$-type ($ErCl_3$, $HoCl_3$, $DyCl_3$). The corresponding entropy range are 50 ± 4, 40.9 and 31 ± 4 J mol^{-1}K^{-1}, respectively.

The (transition + fusion) contribution to entropy ($\bullet_{trs}S_m + \bullet_{fus}S_m$) calculated from data obtained separately [23] on $TbCl_3$ ($PuBr_3$-type structure) and $ErCl_3$ ($AlCl_3$-type structure) is in excellent agreement with previous observations. Moreover, the entropy of fusion values $\bullet_{fus}S_m$ of [23] and $EuCl_3$ (56.3 J mol^{-1}K^{-1}) suggest that both chlorides have the UCl_3-type structure, while $TmCl_3$ has the $AlCl_3$-type structure. This is in excellent agreement with literature data: $SmCl_3$ and $EuCl_3$ have the UCl_3-type structure [24] and $TmCl_3$ has the $AlCl_3$-type structure [25].

Neutron diffraction investigation of UCL3

Experimental

The neutron diffractions were conducted at the Institute Laue Langevin (ILL), Grenoble (France). The scattering intensities range was in 2θ from 1.45° to 143.6°, i.e. $0.23 \leq Q$ (Å^{-1}) ≤ 16.95. UCl_3 melt was enclosed in silica containers inside vanadium furnace at 1 183 K.

Computational procedure

Two approaches were used [6]:

- The Rigid Ion Model (RIM) with a Born-Mayer like potential.

- The Polarisable Ion Model (PIM) which takes into account the polarisation effects.

Structural results

The total structure factor, $F(Q)$ for molten UCl_3 obtained by ND and MD are shown in Figure 5. The total RDF, $G(r)$ obtained by Fourier transformation of the experimental $F(Q)$ are shown in Figure 6 along with neutron weighted RIM and PIM simulation results and the XRD results by Okamoto, *et al.* [26].

The partial pdfs, $g_{\alpha\beta}(r)$ obtained from PIM simulations for molten UCl_3 (Figure 7) were compared with those in molten $DyCl_3$ [2,3]. The running co-ordination numbers obtained from the three partial $g_{\alpha\beta}(r)$s are shown in Figure 8. A comparison was also performed between the bond-angle distributions, Cl-M-Cl (M = U and Dy) obtained from the PIM and RIM simulations for molten UCl_3 and $DyCl_3$.

Conclusions

- Neutron diffraction, computer simulations and thermodynamic studies have revealed that UCl_3 resembles in its physico-chemical, thermodynamic and structural behaviour to lighter rare earth trichlorides.

- Contrary to earlier X-ray and MD simulations, which showed U^{3+} to be six-fold co-ordinated, the present studies reveal that the cation in molten UCl_3 is seven-fold co-ordinated. Thus, contrary to earlier reports, the co-ordination number in UCl_3, similar to that reported recently in $LaCl_3$, does not change substantially on melting.

- This is also in contrast to the structure of molten $DyCl_3$, which we investigated earlier, where the octahedral arrangement of Cl^- around Dy^{3+} is more stable and predominant. In the molten state, UCl_3 is found to be structurally isomorphic to $LaCl_3$, and $DyCl_3$ to YCl_3.

- It is inferred that the six-fold co-ordination becomes progressively more stable with a decrease in cation radius from La (or U) to Dy (or Y) across the lanthanide chloride series.

- The UCl_3 melt shows a pre-peak (FSDP) at $Q \sim 1$ $Å^{-1}$ in the structure factor data. This pre-peak, regarded as the signature of IRO, arises due to various types of connectivity between local co-ordination polyhedra.

Thermodynamics of NaCl-UCl₃

The enthalpy of mixing of the $NaCl-UCl_3$ system was recently measured by direct high temperature calorimetry [27]. The experimental technique has been described earlier [28].

A preliminary assessment of the available thermodynamic data was performed using the Thermocalc software. The phase diagram of the NaCl-UCl₃ has been reported in the "Reactor Handbook Materials" [29] with a simple eutectic at T = 793 K and x(UCl₃) = 0.32. However, Kraus' earlier work [30], which was then confirmed by Barton, *et al.* [31], quoted T = 798 K and x(UCl₃) = 0.33 for the eutectic co-ordinates. More recently, a lower eutectic temperature T = 781 K, at x(UCl₃) = 0.32, was determined by Desyatnik, *et al.* [32].

These data together with the present enthalpy-of-mixing data for the mixture were used in the optimisation procedure with Thermocalc. The input data for UCl₃ were those included in the Thermocalc database. The ionic two-sublattice liquid model was used for modelling the liquid phase. Figures 9-11 show the results of this computation, which include an estimation of activities. The calculated enthalpy of mixing (Figure 9) and phase diagram (Figure 10) fit very well the experimental values [27] and [32], respectively. As no experimental data are available in literature, activities of NaCl and UCl₃ in the melt were calculated (Figure 11).

REFERENCES

[1] M. Gaune-Escard, L. Rycerz, "Molten Salt Chemistry and Technology 5", Molten Salt Forum (ISSN 1021-6138), Vol. 5-6 (1998), 217-222, H. Wendt, ed., Trans. Tech. Publications, Switzerland.

[2] M. Sakurai, A.K. Adya, R. Takagi, M. Gaune-Escard, *Z. Naturforsch.*, Part a: 53a, 655-658, (1998).

[3] A.K. Adya, R. Takagi, M. Gaune-Escard, *Z. Naturforsch.*, 53a, 1037-1048, (1998).

[4] L. Rycerz, M. Gaune-Escard, *High Temp. Material Processes*, 2, N° 4, 483-496 (1998).

[5] L. Rycerz, M. Gaune-Escard, *Z. Naturforsch.*, 54a, 229-235, (1999).

[6] R. Takagi, F. Hutchinson, P.A. Madden, A.K.Adya, M. Gaune-Escard, *Physics: Condensed Matter.*, 11, 645-658 (1999).

[7] L. Rycerz, M. Gaune-Escard, *J. Thermal Analysis and Calorimetry*, 56, 355-363 (1999).

[8] M. Gaune-Escard, L. Rycerz, M. Hoch, *J. Molecular Liquids*, 83, 83-94 (1999).

[9a] M. Gaune-Escard, F. Da Silva, L. Rycerz, R. Takagi, Y. Iwadate, A. Adya, ECS Meeting Abstracts, Vol. MA 98-1, 1998, p. 1102, 11th International Symposium on Molten Salts, P.C. Trulove, H.C. De Long, G.R. Stafford and S. Deki, eds., PV 98-11, San Diego, California, Spring 1998, hardbound, ISBN 1-56677-205-2.

[9b] M. Gaune-Escard, Y. Koyama, R. Takagi, K. Fukushima, Y. Iwadate, *J. Molecular Liquids*, 83, 105-110 (1999).

[10] M. Gaune-Escard, A. Bogacz, L. Rycerz, W. Szczepaniak, *J. Alloys Comp.*, 235, 176 (1996).

[11] M. Gaune-Escard, L. Rycerz, W. Szczepaniak, *J. Alloys Comp.*, 204, 193 (1994).

[12] N.M. Kulagin, D.M. Laptev, *Russ. J. Phys. Chem.*, 50, 483 (1976).

[13] J.L. Moriarty, *J. Chem. Eng. Data*, 8 (3), 422 (1963).

[14] O.G. Polyachenok, G.I. Novikov, *Russ. J. Inorg. Chem.*, 9 (4), 429 (1964).

[15] D.M. Laptev, N.M. Kulagin, I.S. Astakhova, N.V. Tolstoguzov, *Russ. J. Inorg. Chem.*, 26 (4), 553 (1981).

[16] J.A. Sommers, E.F. Westrum, *J. Chem. Thermodyn.*, 9, 1 (1977).

[17] Z.B. Pankratz, *Thermodynamic Properties of Halides*, Bull. 674, 1984 (US Bureau of Mines).

[18] I. Barin, O. Knacke, O. Kubaschewski, "Thermochemical Properties of Inorganic Substances, Supplement", Springer-Verlag Berlin, Heidelberg, New York, 1977.

[19] O. Knacke, O. Kubaschewski, K. Hesselmann, "Thermochemical Properties of Inorganic Substances", Springer-Verlag Berlin, Second Edition, 1991.

[20] O. Kubaschewski, C.B. Alcock, "Metallurgical Thermochemistry", 5th ed., Oxford, New York, Toronto, Sydney, Paris, Frankfurt, 1979.

[21] A.S. Dworkin, M.A. Bredig, *High Temp. Sci.*, 3, 1, 81 (1971).

[22] M.P. Tosi, G. Pastore, M.L. Saboungi, D.L. Price, *Physica Scripta*, T39, 367 (1991).

[23] L. Rycerz, M. Gaune-Escard (unpublished work).

[24] J.A. Gibson, J.F. Miller, P.S. Kennedy, G.W. Prengstorff, "The Properties of the Rare Earth Metals and Compounds", compiled for The Rare Earth Research Group (1959).

[25] D.H. Templeton, G.F. Carter, *J. Phys. Chem.*, 58, 940 (1954).

[26] Y. Okamoto, F. Kobayashi and T. Ogawa, *J.Alloys Comp.*, 271-273 (355) (1998).

[27] H. Matsuura, R. Takagi, L. Rycerz, M. Gaune-Escard (to be published).

[28] M. Gaune-Escard, A. Bogacz, L. Rycerz, W. Szczepaniak, *Thermochimica Acta*, 236, 67-80 (1994).

[29] C.R. Tipton, "Reactor Handbook Materials", 2nd edition, NY (1960), p. 442.

[30] C.A. Kraus, report N° M-251, 1 July (1943).

[31] C.J. Barton, *et al.* in "Phase Diagrams of Nuclear Reactors Materials", R.F. Thoma, ed., ORNL-2548, 133 (1959).

[32] V.N. Desyatnik, B.N. Dubinin, S.P. Raspopin, *Zh. Fiz. Kim*, 47, 2726 (1973).

Figure 1. Heating thermogram of EuCl₃ (1st run)

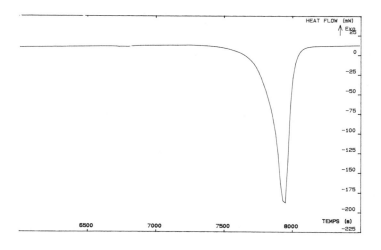

Figure 2. Cooling thermogram of EuCl₃ (1st run)

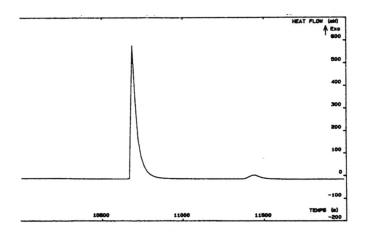

Figure 3. Heating thermogram of EuCl₃ (2nd run)

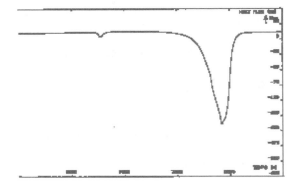

Figure 4. Molar heat capacity of EuCl₃

O – *experimental (this work)*
● *– Sommers and Westrum* [16]
Dashed line 1 *– polynomial fit of exp. data*
Full line 2 *– Pankratz's estimate* [17]

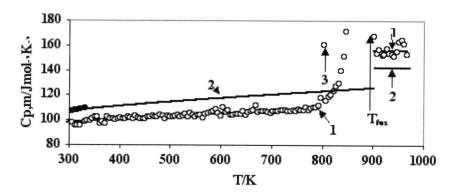

Figure 5. Total structure factor, $F(Q)$

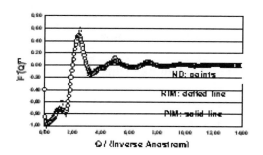

Figure 6. Total RDF, $G(r)$

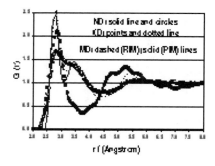

Figure 7. Partial pdfs, $g_{\alpha\beta}(r)$

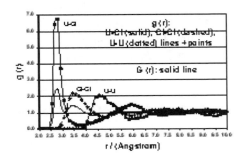

Figure 8. Running co-ordination numbers

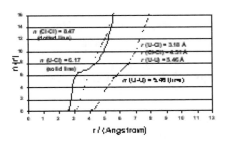

Figure 9. Experimental [27] and calculated (this work) enthalpy of mixing of NaCl-UCl₃

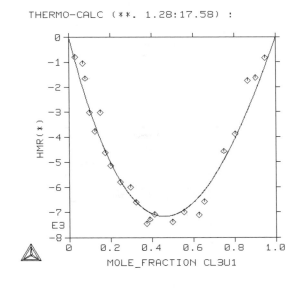

Figure 10. Experimental [32] and calculated phase diagram of NaCl-UCl₃ (this work)

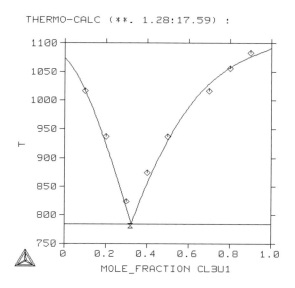

Figure 11. Calculated activities of NaCl (1) and UCl₃ (2) (this work)

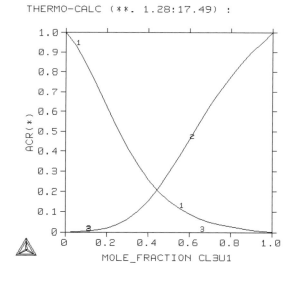

ELECTROCHEMICAL PROPERTIES OF PLUTONIUM IN MOLTEN CaCL2-NaCL AT 550°C

David Lambertin, Jérôme Lacquement
CEA Marcoule
DCC/DRRV/SPHA/LEPP
BP 171, 30207 Bagnols sur Cèze, France

Sylvie Sanchez, Gérard Picard
Laboratoire d'Electrochimie et de Chimie Analytique
(UMR 7575 du CNRS) ENSCP
11, rue Pierre et Marie Curie, 75231 Paris, France

Abstract

The chemical and electrochemical properties of plutonium chlorides have been studied in the fused CaCl2-NaCl equimolar mixture at 550°C using a tungsten working electrode and a pO^{2-} indicator electrode. The standard potential of Pu(III)/Pu was determined using both cyclic voltammetry and convoluted curves. The solubility product of Pu_2O_3 was calculated by potentiometric titration. All these data allowed us to draw the potential pO^{2-} diagram which summarises the properties of plutonium compounds in the melt.

Introduction

Pyrochemical techniques, which consist in performing high temperature separations in a molten salt medium, appear as a promising and valuable route compared to aqueous methods in the field of the separation and transmutation strategies for the long life elements (1991 French law) [1]. The available molten salts, suitable for chemical processes, are numerous. A good knowledge of molten salt actinide chemistry (and in particular for plutonium and americium) is essential to master those separation techniques. Hence, obtaining elementary data in molten halogenides salts is a major concern. Many studies have been conducted on the plutonium in molten chloride salt (in particular in LiCl-KCl) by EMF measurement, chronopotentiometry and voltammetry [2-5]. However comparison of these results is difficult because of the variety of experimental conditions (temperature, reference electrode, composition of melt, purification).

A salt used in plutonium purification [6] has been employed to determine plutonium behaviour. Actinide behaviour in molten salt depends both on potential and oxoacidity level, and these two parameters should be taken into account. Thus in order to describe actinide behaviour in molten chlorides in a summarising form, potential oxoacidity (pO^{2-}) diagrams are necessary. In fact, these diagrams give an instantaneous and comprehensive view of the properties of selected elements in a solvent of interest.

Experimental

Apparatus and electrodes

The equimolar mixture ($CaCl_2$ and $NaCl$, ACS reagents) is melted in a vitreous carbon crucible placed in a quartz cell inside a furnace. The temperature of the furnace is controlled by a Pekly XS 30 programmable device. The mixture is fused under atmospheric pressure using dry argon. No preliminary purification is performed. The pO^{2-} indicator electrode is a tube of yttria-stabilised zirconia, supplied by Degussa (inner diameter 4 mm; outer diameter 6 mm), filled with the molten salt $CaCl_2$-$NaCl$ containing oxide and silver ions ([AgCl] = 0.75 mol/kg; [Na_2CO_3] = 0.1 mol/kg) in which a silver wire is immersed. Our working electrochemical cell is displayed in Figure 1. The reference electrode is made of a silver wire (1 mm diameter) dipped into a Pyrex tube containing a solution of silver chloride in $CaCl_2$-$NaCl$ (0.75 mol/kg). The 1 mm diameter tungsten wires are used as working and counter electrodes. Electrode active surface area is determined measuring the depth of immersion. Cyclic voltammograms are performed using a Autolab PGSTAT30 potentiostat coupled with a PC computer.

Preparation of plutonium chloride

Plutonium chloride is prepared by Pu/Ga metal oxidation. The first step consists of the oxidation of the Pu/Ga metal with $Cl_{2(g)}$. The Gallium trichloride so obtained is volatile and observed at the top of the cell [11]. After the oxidation step, insoluble plutonium is chlorinated by a carbo-chlorination: graphite powder is added in the melt and $Cl_{2(g)}$ is bubbled through the melt using a graphite gas tube. The molten salt takes on a light green tinge indicating $PuCl_3$ formation. The concentration of $PuCl_3$ is determined by alpha-analysis of salt dissolved in nitric acid.

Results and discussion

Electrochemistry of plutonium

The electrochemistry of Pu is studied at a metallic tungsten electrode. Cyclic voltammograms are performed at different sweep rates on a tungsten electrode (Figure 2). Many studies have demonstrated the Pu(III)/Pu system in molten chlorides [2-5,8]. Our study shows that the peak potential is not sweep rate dependent, indicating the reversible character of Pu(III)/Pu redox system. In this case, plots of the peak current versus the square root of the sweep rate is found to be linear (Figure 3), characterising a diffusionnal control of the mechanism. In the LiCl-KCl melt the reversibility of plutonium system was not clearly shown. Burris [9] reported that the plutonium system is reversible and controlled by the diffusion mass transfer, whereas Martinot and Duyckaerts [10] described the plutonium system as an irreversible redox system. In the eutectic NaCl-CaCl₂ melt fused at 550°C, the plutonium redox system appears as a reversible system controlled by a diffusion step.

Standard potential of Pu(III)/Pu determination

In order to determine the standard potential of Pu(III)/Pu(0) reversible redox system, we performed the semi-integration of the voltammograms according to the convolution principle [12,13]. The main advantage of this method is a best accuracy because the mathematical treatment is applied along the whole I-E curve and does not use only the peak values. According to a soluble-insoluble reversible system limiting by a diffusion step, the relationship applied is:

$$E = E_{1/2} + \frac{RT}{zF} \mathrm{Ln}\left(m(t)_{max} - m(t)\right) \tag{1}$$

$$E_{1/2} = E^0 + \frac{2.3RT}{zF} \mathrm{Log}\left(\frac{Pu^{3+}}{2}\right) \tag{2}$$

in which $m(t)$ is the convoluted current (A), E^0 the standard potential of the Pu(III)/Pu system, z the number of electrons exchanged, R the ideal gas constant, T the temperature, F the Faraday constant and Pu^{3+} is the concentration of plutonium trichloride (mol/cm³).

Figure 4 exhibits the cyclic voltammogram and its corresponding convoluted curve calculated using the previous relation (1). The plot Ln(A) (with A = $(m(t)_{max}-m(t))$) versus potential is given Figure 4. The slope is characteristic of a three-electron exchange. The half potential $E_{1/2}$ determined by the analysis of the plot (Figure 4) is equal to -2.108 V versus AgC/Cl taking into account plutonium trichloride concentration determined by alpha analysis ([Pu(III)] = $2.48.10^{-3}$ mol/kg). The standard potential of Pu(III)/Pu is calculated and given Table 1.

Table 1. Value of standard potential of Pu(III)/Pu in different molten chloride mixture. (reference: Cl₂(1 atm)/Cl⁻, mole fraction scale)

	Inoue, et al. [15]	Skiba, et al. [5]		Leary, et al. [2]	This work
Temp. (K)	LiCl-KCl	LiCl-KCl	NaCl-KCl	LiCl-KCl	NaCl-CaCl₂
773	-2.775 V	-2.721 V		-2.72 V	
823		-2.688 V		-2.687 V	-2.667 V
1 000		-2.572 V	-2.642 V		

Calculation of actinide chlorides activity coefficients in the molten salt can yield information about the complexation of the cations by the melt [16]. The activity coefficients, γ_{PuCl_3}, are calculated comparing the standard free energy (determined using electrochemical methods) with the reference free energy (deduced from thermal measurements). The reference state chosen is the super-cooled liquid actinide chloride. Activity coefficients are calculated using the following equation [7]:

$$\log\left(\gamma_{PuCl_3}\right) = \frac{3FE^0 - \Delta G^*_{PuCl_3}}{2.3RT} \tag{3}$$

The standard free energy of formation $\Delta G^*_{PuCl_3}$ of the super-cooled liquid is determined using the published values of Inoue, *et al.* [15]. Table 2 gives the logarithm of the activity coefficient of plutonium in different molten chloride calculated using the potential values given in Table 1.

Table 2. Logarithm of the activity coefficient of plutonium in different molten chloride using standard potential values of the Table 1

	Inoue, *et al.* [15]	Skiba, *et al.*[5]		Leary, *et al.*[2]	This work
Temp. (K)	LiCl-KCl	LiCl-KCl	NaCl-KCl	LiCl-KCl	NaCl-CaCl$_2$
773	-2.4	-1.35		-1.33	
823		-1.2		-1.18	-0.82
1 000		-0.82	-1.88		

The activity coefficients calculated in molten chloride is characteristic of the complexation of plutonium ions in the salt [16]. Considering the melt oxoacidity (Table 3), it appears that the complexation of plutonium is more important when the equilibrium (I) constant value is larger. The equilibrium constant value characterises the oxoacidity of the melts; the less important the constant value, the more important the complexation of the oxide ions in the melt, and thus the more oxoacid the melt. Therefore, the classification of the molten salts in order to oxoacidity is:

$$CaCl_2\text{-}NaCl > LiCl\text{-}KCl > NaCl\text{-}KCl$$

Table 3. Logarithm of the oxoacidity equilibrium (I) constant of HCl$_{(g)}$/H$_2$O$_{(g)}$ couple (I) in various chloride melts

O^{2-}+2HCl$_{(g)}$⇔H$_2$O$_{(g)}$+2Cl$^-$ (I)			
Temp. (K)	CaCl$_2$-NaCl [17]	LiCl-KCl [18]	NaCl-KCl [18]
773		9.42	
823	5.9	8.64	
1 000		6.49	13.8

The complexation power of the melt depends on the nature of the salt cations and on the working temperature. Using monovalent cations (Li, Na or K) or working at higher temperature increase the complexation phenomenon. But it seems (comparing Tables 2 and 3) that the complexation power of a molten salt can be quantified only by its oxoacidity measurement.

The largest activity coefficient of plutonium is obtained in $CaCl_2$-NaCl at 550°C which is the most oxoacid of the molten salts considered and corresponds to a low complexation by the chlorine ions. That is also characterised by the standard potential value measured by electrochemistry, which is close to the reference value.

Potential pO^{2-} diagram of plutonium

Potential acidity stability ranges of plutonium compounds in the molten salt $CaCl_2$-NaCl can be defined and expressed in equilibrium diagrams potential pO^{2-} [19]. The nature of plutonium oxides as well as their solubility products can be determined by potentiometric titration of an oxoacid (Lux-Flood oxoacidity [20,21]) by an oxide ion. The titration is based on the pO^{2-} (= -log $[O^{2-}]$) measurement using an yttria-stabilised zirconia membrane electrode (YZME) which presents a Nernstian behaviour in this melt [22].

In order to evidence and to determine the stability of plutonium oxide compounds, the titration of Pu^{3+} by O^{2-} ions was realised adding small amounts of sodium carbonate in the molten salt containing an initial concentration of $PuCl_3$ equal to $2.48 \cdot 10^{-3}$ mol/kg. The titration curve given in Figure 5 shows only one equivalent point for α equal to 1.5 (defined as the ratio of added O^{2-} ion over the initial Pu(III) concentration, C_0).

This value characterises the formation of Pu_2O_3 according to the following reaction:

$$2Pu^{3+} + 3O^{2-} \rightarrow Pu_2O_{3(s)} \tag{10}$$

Moreover, in agreement with the thermodynamic data, no oxychloride species were evidenced under our experimental conditions.

The analysis of the titration curve leads to the determination of the solubility product $K_{s(Pu_2O_3)}$. The following mass balance equations can be given:

Plutonium for which initial concentration is noted as C_0:

$$C_0 = [Pu^{3+}] + \left[\frac{2nPu_2O_3}{m}\right] \tag{4}$$

in which nPu_2O_3 is the mole number of Pu_2O_3 and m the melt mass.

For oxide ions, knowing that $\alpha = [O^{2-}]_{added}/C_0$, we obtain:

$$\alpha C_0 = [CO_3^{2-}] + [O^{2-}] + \left[\frac{3nPu_2O_3}{m}\right] \tag{5}$$

The solubility product of reaction (10) is given by:

$$K_{s(Pu_2O_3)} = [Pu^{3+}]^2 * [O^{2-}]^3 \tag{6}$$

The free oxide concentration depends on the following equilibrium:

$$CO_3^{2-} \rightarrow CO_2 + O^{2-}$$

the constant of which is noted as K_d and given using the relation:

$$K_d = \frac{P_{CO_2} * [O^{2-}]}{[CO_3^{2-}]} \tag{7}$$

we can derive the following equation for the titration curve:

$$\alpha = 1.5 + \frac{1}{C_0}\left[[O^{2-}] * \left(\frac{P(CO_2)}{K_d} + 1 \right) - 1.5 * \left(\frac{K_{s(Pu_2O_3)}}{[O^{2-}]^3} \right)^{1/2} \right] \tag{8}$$

The solubility product value $K_{s(Pu_2O_3)}$ was calculated fitting the experimental points obtained during the titration of Pu^{3+} by O^{2-} before the equivalent point. Under these conditions, the two first terms of Eq. (8) not being considerable, the relation is simplified and becomes:

$$pO^{2-} = -\frac{1}{3}\text{Log}\left[\frac{K_{s(Pu_2O_3)}}{C_0^2\left(1 - \frac{2}{3}\alpha\right)^2} \right] \tag{9}$$

Values of pKs obtained from potentiometric titration and thermodynamic calculations are gathered in Table 4. In order to estimate the value of $pK_{s(PuOCl)}$, we used the constants K related to the following equilibrium between pure compounds:

$$PuOCl_{(s)} + 2NaCl_{(l)} \xrightarrow{K_1^*} PuCl_{3(l)} + Na_2O_{(s)}$$

$$Pu_2O_{3(s)} + 6NaCl_{(l)} \xrightarrow{K_2^*} 2PuCl_{3(l)} + 3Na_2O_{(s)}$$

Table 4. Solubility products of plutonium oxides in the equimolar CaCl₂-NaCl at 823 K

Equilibrium	Expression for solubility product	Stability constant pKs (molal scale)
$PuOCl_s \rightarrow Pu^{3+} + O^{2-} + Cl^-$	$K_{s(PuOCl)} = [Pu^{3+}] * [O^{2-}]$	6.08[a]
$Pu_2O_{3(s)} \rightarrow 2Pu^{3+} + 3O^{2-}$	$K_{s(Pu_2O_3)} = [Pu^{3+}]^2 * [O^{2-}]^3$	17.6[b] 17.22[a]

[a] Calculated from thermodynamic data.
[b] Obtained from potentiometric titration.

As shown by Mottot [14], the following relations lead to solubility product determination:

$$pK_{s(PuOCl)} = pK_1^* - 2\log a_{NaCl} + \log\gamma_{Pu^{3+}} + \log\gamma_{Na_2O}$$

$$pK_{s(Pu_2O_3)} = pK_2^* - 6\log a_{NaCl} + \log\gamma_{Pu^{3+}} + 3\log\gamma_{Na_2O}$$

in which K_1^* and K_2^* are calculated using the thermodynamical data of the literature [23]. The activity coefficient of Pu^{3+} was previously determined for this melt (Table 2). The activity coefficient of Na_2O is derived from experiment data and chemical potentials ($\log\gamma(Na_2O) = -18$). NaCl activity a_{NaCl} is 0.27 [22]. The value of pK_s so obtained is given Table 4.

Then, oxychloride is not stable for our experimental conditions. Precipitation of oxychloride, however, depends on the plutonium concentration in the melt. This phenomena is shown Figure 6 which gives the variation of $\log[Pu^{3+}]$ versus pO^{2-} calculated using the solubility product relations given in Table 4.

These results allowed us to draw the potential pO^{2-} equilibrium diagram for different plutonium concentrations (Figure 7). The relations between potential and pO^{2-} are given in Table 5.

Table 5. Relation between potential and pO^{2-}

Redox systems	Expression of the equilibrium potential	Standard potential vs. Cl_2/Cl^- at 1 atm (molal scale) (V)
$Pu(III) + 3e^- \rightarrow Pu_{(s)}$	$E = E^0 + \dfrac{2.3RT}{3F} Log[Pu(III)]$	$E^0 = -2.732$
$PuOCl_{(s)} + 3e^- \rightarrow Pu_{(s)} + O^{2-} + Cl^-$	$E = E^1 + \dfrac{2.3RT}{3F} pO^{2-}$	$E^1 = E^0 - \dfrac{2.3RT}{3F} pK_{s(PuOCl)} = -3.06$
$Pu_2O_{3(s)} + 6e^- \rightarrow 2Pu_{(s)} + 3O^{2-}$	$E = E^2 + \dfrac{2.3RT}{2F} pO^{2-}$	$E^2 = E^0 - \dfrac{2.3RT}{6F} pK_{s(Pu_2O_3)} = -3.21$

Conclusions

This study confirms the stability of two oxidation states of plutonium (0 and III) in the equimolar mixture $CaCl_2$-NaCl fused at 823 K. Pu(IV) was not evidenced. The electrochemical study indicated a reversible behaviour of the redox system Pu(III)/Pu. The standard potential of the redox couple Pu(III)/Pu is determined using convoluted methods. The calculated value obtained is $E^0_{Pu(III)/Pu} = -2.667 \pm 0.02V$ versus the Cl_2/Cl^- electrode system (mole fraction scale). The determination of the activity coefficient of $PuCl_3$ and its comparison in various melts shows clearly the influence of the nature of the molten salt on the complexation of $PuCl_3$ by chlorine anion.

The stability of Pu-O compounds was investigated. The solubility product of Pu_2O_3 was determined by potentiometric titration of Pu(III) ions by sodium carbonate. pK_s of plutonium oxychloride was calculated combining experimental and thermodynamic data. All these data allow us to draw the potential pO^{2-} equilibrium diagram for various plutonium concentrations which summarises the properties of plutonium compounds in this melt.

REFERENCES

[1]   CEA, "Assessment of Pyrochemical Processes for Separation/Transmutation Strategies: Proposed Areas of Research", PG-DRRV/Dir/00-92 (2000).

[2]   G.M. Campbell, J.A. Leary, "Thermodynamic Properties of Pu Compounds From EMF Measurements", Rapport LA-3399 (1965).

[3]   S.P. Fusselman, D.L. Grimmet, R.L. Gay, J.J. Roy, C.L. Krueger, T.S. Storvick, T. Inoue, "Measurement of Standard Potentials of Actinides (U, Np, Pu, Am) in LiCl-KCl Eutectic Salt and Separation of Actinides from Rare Earths by Electrorefining", *J. of Alloys and Compounds*, 271-273, 592-596 (1998).

[4]   L. Martinot, G. Duyckaerts, "Mesure du potentiel standard du couple Pu(III)/Pu dans l'eutectique LiCl-KCl entre 400 et 600°C", *Bull. Soc. Chim. Belges*, 80, 299-303 (1971).

[5]   Martinot, "Handbook on the Physics and Chemistry of the Actinides", A.J. Freeman and C. Keller, eds., Elsevier Science Publishers B.V. (1991).

[6]   G. Bourges, Procédés Pyrochimiques, Atelier Gédéon, Pantin, France, 16-18 September (1998).

[7]   B. Trémillon, "Électrochimie analytique et réactions en solution", Tome 1, Masson (1993).

[8]   O. Shirai, *et al.* "Electrochemical Behavior of Actinide Ions in LiCl-KCl Eutectic Melts", *J. of Alloys and Compounds,* 271-273, 685-688 (1998).

[9]   L. Burris, Proceedings of the Annual AIChe Meeting, Miami, Florida, 2-7 Nov. (1986).

[10]   L. Martinot, G. Duyckaerts, *Anal. Lett.*,4, 1 (1971).

[11]   J. Lacquement, P. Brossard, J. Bourges, B. Sicard, A.G. Osipenko,P.T. Porodnov, O.V. Skiba, A.V. Bychkov, L.G. Babikov, "Conversion of Weapon-Grade Plutonium into Nuclear MOX Fuel by Pyrochemical Methods", Molten Salt Forum, Vols. 5-6 (1998), 533-536, Trans. Tech. Publications LTD.

[12]   A.J. Bard, L.R. Faulkner, "Electrochemical Methods", Wiley, New York (1980).

[13]   J.C. Imbeaux, J.M. Savéant, *J. Electroanal. Chem.*, 44, 169 (1973).

[14]   Y. Mottot, Thèse de l'Université de Paris VI (1986).

[15]   S.P. Fusselman, T.S. Storvick, T. Inoue, N. Takahashi, "Thermodynamic Properties of U, Np, Pu, and Am in Molten LiCl-KCl Eutectic and Liquid Cadmium", *J. Electrochem. Soc.*, Vol. 143, 2487 (1996).

[16] A.S. Kertes, *Actinides Rev.*, 1, 371 (1971).

[17] Y. Castrillejo, *et al.*, *J. Electroanal. Chem.*, 449, 67-80 (1998).

[18] G.S. Picard, "Three-Dimensional Reactivity Diagrams for Reaction Chemistry in Molten Chlorides", Molten Salt Forum, Vol. 1-2, 23-40 (1993).

[19] B. Trémillon, *Pure Appl. Chem.*, 25, 395 (1971).

[20] H.Z. Lux, *Elektrochemie*, 45, 303 (1939).

[21] H. Flood, "The Acidic and Basic Properties of Oxides", *Acta Chem. Scand.*, 1, 592 (1947).

[22] J. Lumsden, "Thermodynamics of Molten Salts Mixtures", Academic Press, New York (1966).

[23] J. Barin, O. Knacke, "Thermochemical Properties of Inorganic Substances", Springer, Berlin (1973).

Figure 1. Electrochemical cell

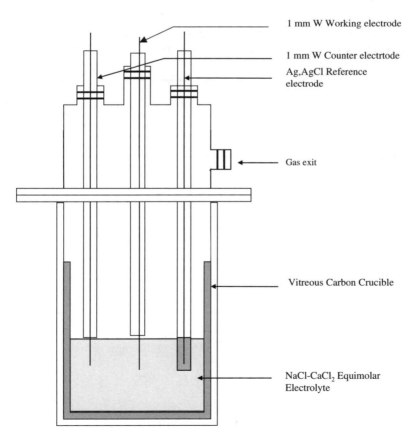

Figure 2. Cyclic voltammograms of PuCl₃ in the molten salt NaCl-CaCl₂ at 823 K

[PuCl₃] = 4.10⁻³ mol/kg
Working electrode: tungsten, Electrode area: 0.604 cm²
Scan rate: 0.2 V/s, 0.25 V/s, 0.3 V/s, 0.4 V/s

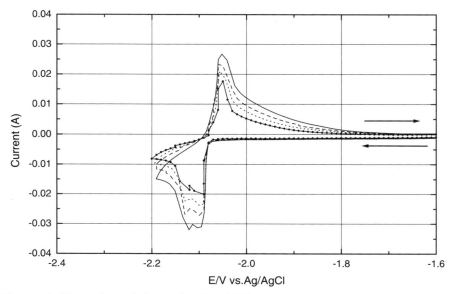

Figure 3. Variation of the cathodic current peak reported from the cyclic voltammograms given in Figure 2 with the square root of the sweep

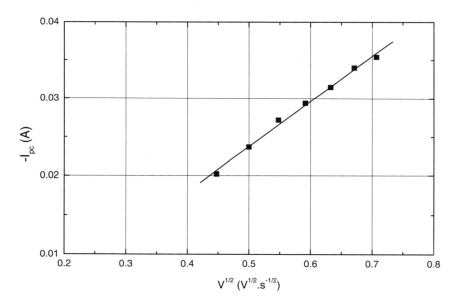

Figure 4. Cyclic voltammogram and its corresponding convoluted curve and the logarithmic analysis of the convoluted curve considering a soluble-insoluble redox system; sweep rate = 0.1 V/s

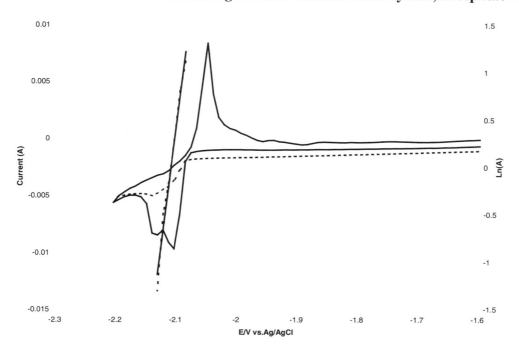

Figure 5. Potentiometric titration curve of 0.00278 mol/kg Pu^{3+} solution by O^{2-} ions added as solid Na$_2$CO$_3$. Experimental points and fit curve.

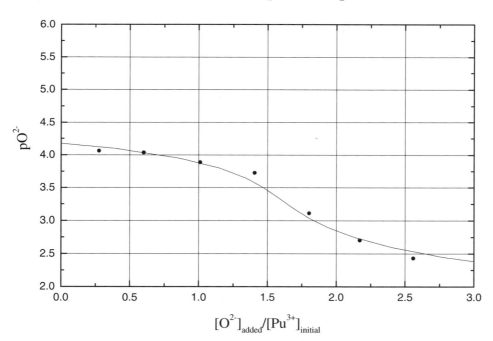

Figure 6. Influence of the concentration of plutonium trichloride on the precipitation of plutonium oxide compounds (molal scale)

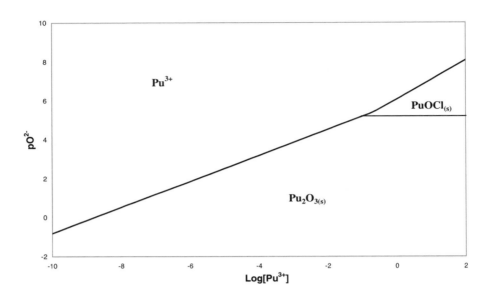

Figure 7. Potential pO²⁻ diagram for plutonium at various concentrations in the equimolar CaCl₂-NaCl mixture at 823 K

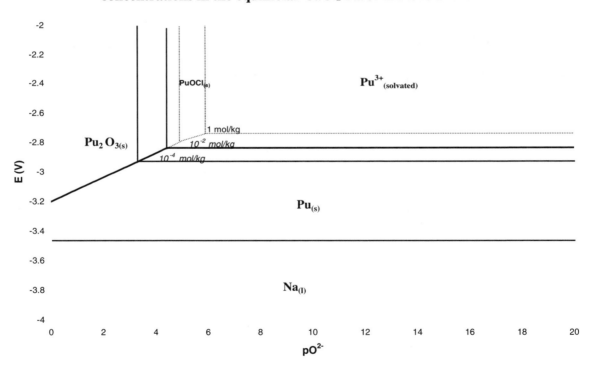

APPLICATION OF ELECTROCHEMICAL TECHNIQUES IN PYROPARTITIONING PROCESSES AND REACTION MODELLING

Frédéric Lantelme, Thierry Cartailler and Pierre Turq
LI2C-Electrochimie, UPMC, 4 place Jussieu, 75252 Paris Cedex 05, France

Abstract

The redox reactions involving rare earth solutions in fused salts were studied by transient electrochemical techniques. Digital simulation was used to obtain a precise representation of the transient responses including perturbing factors such as ohmic drop, electrocrystallisation and reduction of alkali ions. A fitting procedure, included in the numerical calculation, allows an accurate determination of reaction parameters, diffusion coefficients, standard potentials and covering factor. The technique was extended to the formation of alloys with nickel. Examples of application to lanthanum and gadolinium deposition are presented.

Introduction

The design of a metal molten salt based chemical and electrochemical process for treatment of nuclear waste requires reliable basic data [1,2]. The electrochemical techniques provide an efficient tool to investigate the reaction mechanisms. Indeed, the energetic level and the rate of reactions are easily measured from current and voltage. These quantities can be followed all along the progress of the chemical transformation, which gives access to the kinetic parameters and open the way to reaction study by relaxation or transient techniques. These challenging techniques are particularly useful in the field of high temperature chemistry where the reactions are rapid and their use has expended rapidly over the last two decades.

The establishment of modern electrochemical technologies requires early engineering evaluation of all the cell behaviour with reliable procedures. This results in an increasing use of mathematical tools to evaluate the cell and reactor design. Increasingly, these are being used in conjunction with experimental studies of fundamental and laboratory scale operations. The analysis of electrode reactions composed of diffusion, electron transfer, kinetics and additionally adsorption or crystallisation is only possible after suitable models have been formulated [3]. In the next section a rapid presentation of the calculation procedure is given. Afterwards, experiments concerning deposition of lanthanum and gadolinium are described.

Numerical calculation

To draw reliable deductions from the experimental results it is needed to consider the various elemental phenomena which influence the reaction process. Such phenomena include the diffusion of electroactive species, charge transfer reaction, the ohmic drop and the partial coverage of the electrode when a solid compound is deposited at the electrode surface. It is not possible to obtain mathematical analysis of the transient response taking into account these various factors. Digital simulation is the only way to treat the phenomena comprehensively. This consists of solving the partial differential equations of transport by numerical calculation taking into account the boundary conditions coming from the experimental constraints (controlled current or potential) and from the reaction kinetics [4]. A numerical model is set up within a digital computer and the model is allowed to evolve by a set of algebraic laws deriving from equations defining the mass and charge transfer. The space (x co-ordinate) is divided into small intervals of length, Δx, and the time into small time steps, Δt. Concentrations as a function of time and concentration profiles of a species i in the diffusion layer near the electrode were calculated from the diffusion equation:

$$\frac{dc_i}{dt} = D_i \frac{d^2 c_i}{dx^2} \tag{1}$$

In finite difference schemes the continuous functions are finitised with centred finite difference for space derivative. The diffusion flux of the electroactive species i is related to the concentration gradient at the electrode surface according to:

$$J_i = n_i F D_i \frac{c_i(0,t) - c_i(\Delta x, t)}{\Delta x} \tag{2}$$

$c_i(0,t)$ is the concentration at the electrode surface, $c_i(\Delta x,t)$ is the concentration at the distance, Δx. Two reactions are involved in the electrochemical process:

$$Ln^{3+} + 3e^- \leftrightarrow Ln \tag{3}$$

$$M^+ + e^- \leftrightarrow M \tag{4}$$

Ln is the lanthanide metal and M represents the alkali metal. According to the experimental results, the two above reactions are reversible. The concentration of dissolved alkali metal obeys the Nernst equation:

$$c_M(0,t) = \exp\left[-F\left(E(0,t) - E_M^0\right)/RT\right] \tag{5}$$

E_M^0 is the standard potential for reactions (4) i.e. the potential of alkali metal deposition in the electrolyte, the concentration of M^+ remains fixed since the alkali salt is a component of the electrolyte. The surface concentration of lanthanide ions is:

$$c_{Ln^{3+}}(0,t) = \exp\left[3F\left(E(0,t) - E_{3/0}^0\right)/RT\right] \tag{6}$$

$E_{3/0}^0$ is the standard potential for reactions (3). However Eq. (6) is valid when reaction (4) occurs, i.e. for the reduction process or when the rare earth metal covers the electrode surface. During the oxidation step it may occur that the electrode is partly covered by the metal. On the covered part the surface concentration is calculated from Eq. (6). Since no reaction occurs on the uncovered part, the flux $J_{Ln^{3+}}$ is null and the surface concentration is equal to the concentration in the first calculation layer at a distance Δx. According to an hypothesis already used in literature the mean surface concentration is given by:

$$c_{Ln^{3+}}(0,t) = q\exp\left[3F\left(E(0,t) - E_{3/0}^0\right)/RT\right] + (1-q)c_{Ln^{3+}}(Dx,t) \tag{7}$$

The value of the covering factor, θ, is calculated from the number of moles, Q, of metal deposited at the electrode surface:

$$\theta = \frac{Q}{Q_T} \tag{8}$$

Q_T is a parameter which represents the number of moles of deposit required to obtain a full coverage of the electrode surface. Of course, the value of θ is limited to unity, $1 \geq \theta \geq 0$.

The diffusion fluxes $J_{Ln^{3+}}$ and J_M obey Eq. (2). The set of equations are solved by numerical calculation. The total current density, j at each step is given by:

$$j = F(3J_{Ln^{3+}} - J_M) \tag{9}$$

At each step the contribution of the perturbing phenomena was included in the calculation. For example, the real potential, E(t), at the electrode surface was calculated from the recorded overpotential, E′(t), taking into account the contribution of the ohmic drop:

$$E(t) = E'(t) - Ri(t) \tag{10}$$

i(t) is the experimental current at the time, t, and R is the ohmic resistance of the electrical circuit. For more details concerning the calculation processes see our previous publications [3-6]. Our purpose is now to study the deposition of rare earths from fused salt electrochemistry; the experimental results will be interpreted in the frame of a numerical model.

Experimental

Apparatus

The essential component of the cell was a sealed Pyrex container resistant to a vacuum of 0.1 Torr at high temperature (Figure 1). The cell was maintained at a desired temperature in a specially built transparent furnace, made of a nichrome ribbon twisted around a silica glass cylinder [7]. The bath temperature was measured with a calibrated chromel-alumel thermocouple (precision: (±0.5°C)). A PID controller PYREG was used to obtain a good constant temperature. The handling of the fused salt required the use of suitable atmospheres; special attention was devoted to the gas circuit [7] (Figure 2).

Materials

Electrolyte

The electrolyte was a mixture of 58.2 mol% of LiCl and 41.8 mol% of KCl. Before each measurement the salt mixture was dried under vacuum at 120°C, and then heated to 400°C under HCl gas for 1 h. The measurements were performed under argon atmosphere.

Solute

The metal ions were introduced into the molten mixture by anodic dissolution of pure metal rods. In order to remove any traces of oxide ions the solution is treated by chlorine bubbling just before the experimental determination. Of course, during the bubbling the metallic electrodes were removed from the cell. The concentration of the lanthanum or gadolinium was determined coulometrically and chemically.

Electrodes

Reference electrode

A chlorine electrode [8] was used as a reference electrode. The reduction of the Cl_2 gas occurs at the three-phase boundary: solid electrode-liquid electrolyte-gaseous reagent. In molten chlorides such an electrode represents a stable and precise reference electrode. According to Ref. [9] the standard potential of the Cl_2 electrode versus Pt/Pt^{2+} is +0.216 V at 450°C (mole fraction scale).

Working electrode

The thermodynamic measurements were carried out with a two-electrode device, the EMF measurement being carried out between a chlorine electrode and a metallic electrode made of a lanthanum or gadolinium rod (6.35 mm diameter rod, Goodfellow Cambridge Limited, purity of

99.9%). A three-electrode device was used to perform the transient techniques. The working electrode was a tungsten rod (Johnson Matthey, diameter 1.36 mm), the counter electrode was a lanthanum or gadolinium rod. The active area of the working electrode was calculated from the immersion length measured through the transparent window with a cathetometer. The result was controlled by visual observation of the surface wetted by the salt when the electrode was removed from the bath. In some experiments, variable immersion lengths were used. Assuming that the surface increment was well known, the corresponding changes in the transient response allows the accurate determination of the transport parameters [10]. The ohmic resistance of the electrical circuit was determined by the current interruption technique [11]. The resistance ranged between 0.1 to 0.3 Ω.

Anodic dissolution

A two-electrode system was used. The anode was a rod of the metal to be oxidised; the cathode was a chlorine electrode which generates chloride ions.

Instrumentation

For the determination at zero current (EMF), the potential of the working electrode against the Cl_2 reference electrode was measured between to 380 to 590°C at each metal ion concentration using an electronic voltmeter, Keithley 195A, the actual precision of the method being about ±0.5 mV. The transient techniques were carried out using a Tacussel PRT 20-2X potentiostat. A device with a calibrated resistor was used to obtain a galvanostatic control. The generation of different potential or current programmes was carried out with a Tacussel GSTP4 programmer. The transient responses were collected on a Sefram X/Y recorder, or stored temporarily in a Nicolet 310 digital oscilloscope.

Metal deposition

Transient techniques

The mechanism of electroreduction of rare earth ions in fused salt were studied by transient electrochemical techniques. In order to obtain the deposition of pure metal, an inert working electrode must be used. Preliminary experiments showed that refractory metals such as tungsten or molybdenum are convenient electrode materials. The transient electrochemical techniques were used to determine the reaction mechanism. As mentioned in the experimental section, the Cl_2/Cl^- electrode has been applied as a reference electrode.

Voltammetry

Figure 3 shows typical voltammograms at 475°C of an eutectic LiCl-KCl melt with $LaCl_3$. A reduction current peak is obtained at a potential, E_p = -3.235 V. This part of the voltammogram does not depend on the limits of the potential window. An increase of the reduction current is seen in the region of the most negative potentials. During the reverse sweep a steep peak is obtained at -3.18 V. The height of the peak depends on the value of the potential limit (left-hand side): the more negative the potential, the higher the peak. Moreover, the increase of the peak height induces a small positive shift of the peak potential. Over the potential range, -3.15, -3.0 V, the current remains quite small, less than 6 mA. Additional experiments showed that the values of the current and of the peak potential depend on the sweep rate. The current for the reduction peak was proportional to the square root of the sweep rate. At high sweep rates the oxidation peak became broader. The voltammograms obtained with a solution of GdCl₃ behaved in a similar way, with the exception that a positive shift in the value of the peak potentials occurred (Figure 4).

Chronopotentiometry

The analysis by chronopotentiometry of a solution of LaCl₃ showed the existence of a potential plateau in the potential range -3.25, -3.3 V (Figure 5). After the plateau there is a rapid decrease in the potential. The time at which the inflexion point occurs is called transition time, τ. When the constant current was maintained a time longer than the transition time the electrode potential reaches a limit value corresponding to the deposition of alkali metals at about -3.6 V.

The chronopotentiograms obtained with a solution of GdCl₃ behaved in a similar way except for the small positive shift in the value of potential plateau (Figure 6).

Analysis of the experimental results

The mathematical analysis of the voltammograms for electrodeposition of an insoluble product has been investigated by Schiffrin [12] for a diffusion controlled reaction. However, to obtain a comprehensive analysis of the experimental curves some perturbing phenomena must be taken into account, such as the ohmic drop, the partial coverage of the electrode by the metallic deposit and the reduction of alkali metal ions.

Digital simulation was used to exploit the results of transient electrochemical techniques. The values of the fundamental quantities (standard potential, diffusion coefficient, coverage Q_T) are adjusted in order to obtain the best representation of the experimental curves. Figures 3-6 show that the calculated voltammograms and chronopotentiograms correctly fit with the experimental results. According to our model, this agreement indicates that the electrochemical reaction is rapid. The shift in the peak potential when the sweep rate was increased was mainly due to the ohmic drop. The values of the diffusion coefficients were deduced from around thirty transient curves carried out in the temperature interval 380-590°C. In most of the experiments the concentration of electroactive species ranged from 0.2 to 1 mol%. The standard potentials depend slightly on temperature. In the molarity scale it is found:

$$E^0_{La} = -3.568 + 6.1\,10^{-4}\,T \qquad\qquad E^0_{Gd} = -3.511 + 6.5\,10^{-4}\,T$$

The diffusion coefficient obeys an Arrhenius equation:

$$D = D^0 \exp(\Delta H / RT) \tag{11}$$

$D^0_{La^{3+}} = 2.15 \times 10^{-3}\ cm^2\ s^{-1}$ $\qquad\qquad$ $D^0_{Gd^{3+}} = 1.79 \times 10^{-3}\ cm^2\ s^{-1}$

$\Delta H_{La^{3+}} = -30.0\ kJ\ mol^{-1}$ $\qquad\qquad$ $\Delta H_{Gd^{3+}} = -33.0\ kJ\ mol^{-1}$

with a standard deviation $\Delta D = \pm 0.2 \times 10^{-5}\ cm^2\ s^{-1}$. As usual, the ion with the smaller ionic radius diffuses slower ($r_{La^{+++}} = 0.102$ nm, $r_{Gd^{+++}} = 0.094$ nm) which is due the influence of the electrostatic interaction between the cations and the neighbouring chloride ions. The diffusion coefficient of Gd^{3+} has recently been measured by Iizuka [10] who found, $D^0_{Gd^{3+}} = 1.672 \times 10^{-3}\ cm^2\ s^{-1}$ and $\Delta H_{Gd^{3+}} = -31.96\ kJ\ mol^{-1}$.

The mean value of the coverage quantity, Q_T, at 450°C is about 4×10^{-7} mol cm^{-2} for both lanthanum and gadolinium. It seems that Q_T decreases slightly when the temperature is increased. At 550°C the mean value is 2×10^{-7} mol cm^{-2}.

Alloy formation

In the previous sections the metal deposition was studied on an inert electrode. However, rare earths form alloys with many metals and it has been shown that the alloying process can conveniently be studied by electrochemical techniques [13]. To illustrate this behaviour a cyclic voltammogram of $LaCl_3$ in LiCl-KCl is shown in Figure 7. The shape of this curve is very different from the curve obtained at a tungsten electrode. Two remarks must be pointed out. Firstly, care should be taken as to the nature of the working electrode, which may alter the transient response of the ionic reduction. Secondly, the analysis of the voltammogram provides interesting information concerning the alloy formation. As shown by the phase diagram of the La-Ni system [14], the many current peaks are related to various nickel-lanthanum alloys.

Chronopotentiometry was used to study the stability range of the alloys. A constant current electrolysis was used to deposit a known amount of lanthanum at the surface of a nickel electrode. Then, the sign of the current was reversed. The recorded chronopotentiogram shows a few potential plateaux corresponding to two-phase equilibria. In order to obtain an accurate determination of the phase composition, electrolysis at constant potential was performed during a few hours and the surface composition was obtained from EDX and concentration profile analysis of the alloy layer. The potentials of the two-phase systems are given in Table 1. The Gibbs energy of formation of alloys formation deduced from these potentials are reported in Table 2. They are in qualitative agreement with the values measured by Rezukhina and Kutsev [15] at higher temperatures.

Table 1. Equilibrium potentials of the two-phase systems at 460°C vs. La^{3+}/La electrode

Ni/LaNi$_5$	LaNi$_5$/LaNi$_{3.5}$	LaNi$_{3.5}$/LaNi$_3$	LaNi$_3$/LaNi$_2$	LaNi$_2$/LaNi$_{1.5}$	LaNi$_{1.5}$/LaNi	LaNi/LaNi$_{0.33}$
0.483 V	0.205 V	0.201 V	0.131 V	0.091 V	0.078 V	0.045 V

Table 2. Gibbs energy of formation of the alloys at 460°C, kJ mol^{-1}

LaNi$_5$	LaNi$_{3.5}$	LaNi$_3$	LaNi$_2$	LaNi$_{1.5}$	LaNi	LaNi$_{0.33}$
-140	-116	-108	-84	-70	-54	-27

The alloy formation provides a convenient means of improving the extraction of lanthanides from the solution since an important depolarising effect occurs and a large variety of alloys can be used. Moreover, according to the phase diagram, at temperatures greater than around 520°C, liquid alloys can be obtained. Then, a continuous extraction of lanthanides can be performed, and a liquid alloy, such as La$_3$Ni, is continuously flowing from the electrode to the bottom of the cell. It is pointed out that the lowest fused point in the Gd-Ni system (635°C) is much higher than for the La-Ni system (485°C).

Conclusion

The success of a pyropartitioning process of rare earths and actinides in fused salts requires an accurate knowledge of the reaction scheme and of the physicochemical properties of the electrolyte solutions. The stability range of the various ionic species were obtained from EMF measurements using reliable reference electrodes, such as chlorine electrode.

To reach a better view of the feasibility of the process, the kinetic parameters of the reaction steps were measured from transient techniques taking into account the contribution of the different redox couples including alkali metals. Special attention was devoted to the crystallisation process generally controlled by the rate of nucleus formation and the diffusion of active species. A comprehensive approach of the reaction mechanism by molecular dynamic simulation is now investigated. Our aim is to study the interactions at the atomic scale from molecular modelling using mean spherical approximation (MSA) or hyper netted chains (HNC) techniques. This research is in progress to obtain a better view of lanthanides and actinides speciation in fused salts

REFERENCES

[1]   S.P. Fusselman, J.J. Roy, D.L. Grimmett, L.F. Grantham, C.L. Krueger, C.R. Nabelek, T.S. Storvick, T. Inoue, T. Hijikata, K. Kinoshita, Y. Sakamura, K. Uozumi, T. Kawai and N. Takahashi, *J. Electrochem. Soc.*, 146, 2573 (1999).

[2]   F. Lantelme and Y. Berghoute, *J. Electrochem. Soc.*, 146, 4137 (1999).

[3]   F. Lantelme, "Computer Simulation of Electrochemical Processes", in Computerized Physical Chemistry of Metallurgy and Material, Q. Zhiyu, X. Zhihong and L. Honglin, eds., p. 180, Metallurgical Industry Press, Beijing (1999), Engl Trans., LI2C, Case 51, UPMC, 4 place Jussieu, Paris.

[4]   A. Salmi, Y. Berghoute and F. Lantelme, *Electrochim. Acta*, 40, 403 (1995).

[5]   F. Lantelme, Y. Berghoute and A. Salmi, *J. Appl. Electrochem.*, 24, 361 (1994).

[6]   F. Lantelme, "Modelling and Simulation in Fused Salts Electrowinning of Metals", in Modelling and Simulation in Metallurgical Engineering and Materials Science, Y. Zongsen, X. Zeqiang and X. Xishan, eds., p. 133, Metallurgical Industry Press, Beijing (1996).

[7]   F. Lantelme, D. Inman, D.G. Lovering, in Molten Salt Techniques, Vol. 2, R.J. Gale and D. Inman, eds., p. 148, Plenum Press, New York (1984).

[8]   F. Lantelme, H. Alexopoulos, O. Haas, *J. Appl. Electrochem.*, 19, 649 (1989).

[9]   H.A. Laitinen, J.W. Pankey, *J. Am. Chem. Soc.*, 81, 1053 (1959).

[10] M. Iizuka, *J. Electrochem. Soc.*, 14, 84 (1998).

[11] W. J. Wruck, R. M. Machado and T. W. Chapman, *J. Electrochem. Soc.*, 134, 539 (1987).

[12] J. Schiffrin, *J. Electroanal. Chem.,* 201, 199 (1986).

[13] G.S. Picard, Y.E. Mottot and B.L. Trémillon, in Proceedings of 4th International Symposium on Molten Salts, M. Blander, D.S. Newman, M-L. Saboungi, G. Mamantov and K. Johnson, eds., PV 84-2, p. 585, The Electrochemical Society Proceedings Series, Pennington, NJ (1984).

[14] W.G. Moffat, "The Handbook of Binary Phase Diagrams", Vol. 4, p. 7/91, Genium Publishing Corporation, New York (1987).

[15] T.N. Rezukhina and S.V. Kutsev, *Russian Journal of Physical Chemistry*, 56, 1 (1982).

Figure 1. Electrochemical cell

A) Firebrick support
B) Alumina crucible
C) Gas bubbling tube
D) Fused electrolyte
E) Tube with fritted glass n° 5
F) Chlorine electrode
G) Pyrex crucible
H) Feeder tube
I) Counter electrode
J) Working electrode
K) Graphite rod
L) Tube with fritted glass n° 5
M) Alumina cover
N) O-ring
O) Thermocouple
P) Pyrex cell header
Q) Torion joint
R) Rotating tube
S) Feeding device

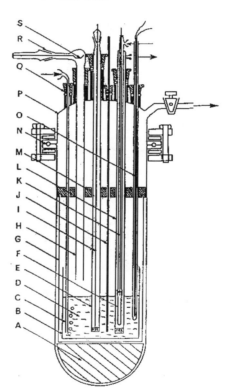

Figure 2. Diagram of the gas circuit

1) vacuum pump; 2) mixing chamber; 3) liquid-air trap; 4) electrochemical cell; 5) neutralisation flask;
6) gas inlet; 7) chlorine electrode; 8) argon cylinder: 9) hydrogen chlorine cylinder; 10) chlorine cylinder

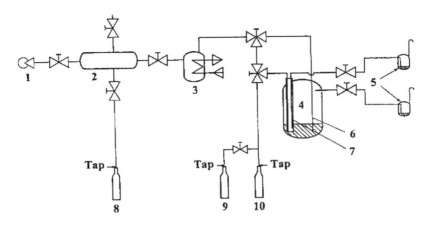

Figure 3. Cyclic voltammogram at 475°C of eutectic LiCl-KCl with LaCl₃ (0.041 mol l⁻¹)

Scan rate: 1 V s⁻¹. Tungsten working electrode (area, 0.39 cm²). Solid line: experimental. Dotted line: calculated.

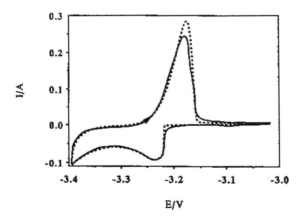

Figure 4. Cyclic voltammogram at 475°C of eutectic LiCl-KCl with GdCl₃ (0.041 mol l⁻¹)

Scan rate: 1 V s⁻¹. Tungsten working electrode (area, 0.39 cm²). Solid line: experimental. Dotted line: calculated.

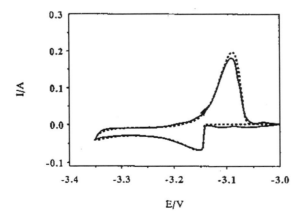

Figure 5. Chronopotentiogram for the reaction
$La^{3+} + 3e^- \rightarrow La$, in LiCl-KCl with $LaCl_3$ (0.0655 mol l^{-1}).

Temperature: 445°C. Current: -50 mA. Tungsten working electrode (area, 0.45 cm².).
Solid line: experimental. Dotted line: calculated.

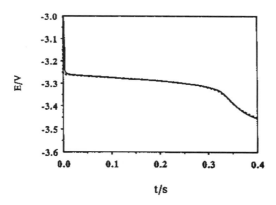

Figure 6. Chronopotentiogram for the reaction
$Gd^{3+} + 3e^- \rightarrow Gd$, in LiCl-KCl with $GdCl_3$ (0.0655 mol l^{-1}).

Temperature: 445°C. Current: -50 mA. Tungsten working electrode (area, 0.45 cm².).
Solid line: experimental. Dotted line: calculated.

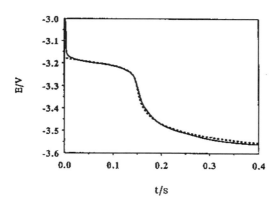

Figure 7. Cyclic voltammogram at 480°C of eutectic LiCl-KCl with $LaCl_3$ (0.1 mol l^{-1})

Scan rate: 2 mV s⁻¹. Nickel working electrode (area, 0.325 cm².).

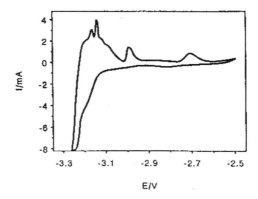

133

Figure 8. Current reversal chronopotentiogram for the reaction La^{3+} + 3e^{-} ↔ La, in LiCl-KCl with LaCl₃ (0.08 mol l^{-1})

Temperature: 460°C. Current: -5 mA; after 140 s: +1 mA. Nickel working electrode (area, 0.35 cm².).

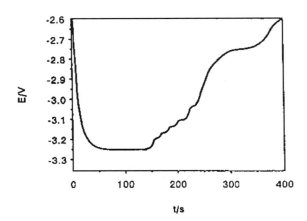

THERMODYNAMIC STABILITIES OF LANTHANIDES AND ACTINIDES IN MOLTEN CHLORIDE AND LIQUID METAL

Hajimu Yamana, Jiawei Sheng and Hirotake Moriyama
Research Reactor Institute, Kyoto University
Kumatori-cho, Sennan-gun, Osaka, 590-0494, Japan
E-mail: yamana@HL.rri.kyoto-u.ac.jp

Abstract

In order to strengthen the comprehensive understanding of the mechanism of the extraction of f-elements in a pyrometallurgical liquid-liquid extraction system, which is usable for the partitioning technique, some thermodynamic quantities related to their affinity to the solvent materials were studied. The distribution characteristics of the f-elements between molten alkaline chloride and liquid Bi or Zn were experimentally studied. The standard Gibbs energy change of formation of diluted liquid alloy of some lanthanides with Bi was experimentally studied, and their systematic variation along the 4-f series was discussed. Based on the results of the extractions, the excess free energies of trichlorides of lanthanides in molten alkaline chloride were estimated, and their systematics were discussed. The standard Gibbs energy change of formation of liquid alloy of actinides with Bi and Zn were estimated from the systematic relation of the thermodynamic quantities.

Introduction

The reductive liquid-liquid extraction of f-elements between molten chloride and liquid metal deserves an examination as a possible technique for the group separation of actinides and lanthanides [1,2]. In a pyrometallurgical liquid-liquid extraction system, the separation performance of elements mainly depends on the standard Gibbs energy of formation of their chlorides. At the same time, however, the activity coefficients of the elements in the metallic and molten salt phase also influence the separation efficiency to a great extent. The thermodynamic activities of the elements in either phase are mostly controlled by their chemical affinity to the solvent. Regarding the liquid metal phase, the metallic states of solute elements form certain intermetallic complexes with the solvent liquid metals causing an excess energetic stabilisation of the solute metals. Thus, the solutes lose their thermodynamic activities from those of the pure states. A similar effect occurs during the cationic states in the molten alkaline chloride salt. Thus, in order to evaluate the effectiveness of the separation of lanthanides and actinides by this technique, their thermodynamic activities in both phases have to be carefully studied.

The thermodynamic activities of metallic lanthanides in liquid metals can be directly determined by an electromotive force (EMF) measurement [3-6], and some of them have been reported as the standard Gibbs energy changes of the solution of f-elements into liquid metals. However, those of metallic actinides are difficult to measure using this technique because of the experimental difficulty in handling radioactive actinides. Regarding the thermodynamic activities of cationic actinides in the molten salt phase, their direct measurement by an electrochemical method is not easy and encounters the same technical difficulty. In addition, due to the difficulty associated with the reference electrodes, reported activity coefficients are likely to scatter.

In this paper, based on the results of the liquid-liquid extraction of f-elements between molten chloride and liquid Bi or Zn, we discuss the systematic characteristics of the thermodynamic stabilities of f-elements in these pyrometallurgical solvents.

Experimentals

Two different experiments, liquid-liquid extraction and electromotive force measurement were performed as described below. For both experiments, a 99.9% pure eutectic mixture of LiCl and KCl (mole ratio of lithium to potassium = 51/49) purchased from Anderson Physics Laboratory Engineered Materials Inc. was used for the molten salt phase. All the extraction experiments were performed in a vacuum tight glove box filled with purified Ar, the humidity and oxygen of which were kept at < 1 ppm. All other reagents used were of analytical grade, and were purchased from Wako Pure Chemicals Co. Ltd. In this study, all the concentrations of the components are given in the unit of a mole fraction.

Liquid-liquid extraction experiments

The experimental procedures to determine the distribution ratios of lanthanides and actinides in a reductive extraction system have been already reported elsewhere [1,2]. Distribution ratios of trivalent lanthanides and actinides were radiochemically determined at 873 and 1 073 K in a bi-phase system of molten chloride and liquid Bi or Zn. By changing the concentration of metallic Li as reductant, the distribution ratio of solutes (D_M) was determined as a function of that of Li (D_{Li}).

Electromotive force determination for lanthanides

In order to reinforce the data of the excess Gibbs energy of lanthanides in liquid bismuth, the electromotive force (EMF) measurement method was applied to the following galvanic cell:

M (solid)/MCl₃ in LiCl + KCl/M(in Bi) where M is Ho, Tb and La

The experimental system placed in an electric furnace consists of a crucible made of finely sintered alumina, thermocouple for temperature measurement, insulated Ta wire for leading the liquid alloy electrode, sampling line and so on. About 36 g of eutectic mixture of LiCl and KCl and 133 g of pure Bi were put in the crucible with about 1 g of one of the lanthanide metals (Ho, Tb and La), then the temperature was raised up to given values by an electric furnace. After the desired temperature was achieved, the bottom tip of the pure lanthanide electrode was immersed into the molten salt phase, and then the difference of the electric potentials between the pure lanthanide electrode and liquid alloy electrode was measured by an electrometer. After immersing the pure lanthanide electrode, the variation of the EMF was monitored for over 10 minutes, waiting for the stabilisation of the EMF. Measurements were performed at around 520, 620, 720 and 820 K for Ho, Tb, and La. At each temperature, the concentration of lanthanide in Bi phase was changed several times by anodic electrolysis or addition of lanthanide pieces. At every concentration after the EMF measurement, a small portion of metallic phase was taken by sucking the solution into a thin stainless pipe, and the mole fraction of lanthanides was analysed by ICP-AES (Shimazu-ICPS-1000III).

The pure lanthanide electrodes used are rods ($3 \times 3 \times 25$ mm) of lanthanide metal of 99.9% purity. Ta wire was welded at one end of the rod. Because there were several cases in which the surface of the electrode was slightly oxidised by traces of oxygen impurity, before every measurement the surface of the pure lanthanide electrode was carefully cleaned with fine sandpaper.

Thermodynamic expression of the extraction reaction

The reductive liquid-liquid extraction of trivalent lanthanides by metallic Li is described by:

$$MCl_3(\text{in S}) + 3Li(\text{in B}) \leftrightarrow M(\text{in B}) + 3LiCl(\text{in S}) \tag{1}$$

where M represents trivalent f-elements, and the salt and metal phase are denoted by S and B, respectively. The brackets mean that the components are dissolved species in designated phases. By adapting the relation between equilibrium constant K and ΔG_{ext}^0, of Eq. (1), Eq. (2) is obtained as a function of the distribution ratios D_M and D_{Li} which are defined by [M(in B)]/[MCl₃(in S)] and [Li(in B)]/[LiCl(in S)], respectively. In Eq. (2), ΔG_{ext}^0 represents the standard Gibbs energy change of Eq. (1).

$$\log\left(D_M / D_{Li^3}\right) = -3\log\gamma_{LiCl} - \frac{1}{2.3RT}\Delta G_{ext}^0 \tag{2}$$

The standard Gibbs energy changes of formation of four species of the reaction are denoted by ΔG_f^0[MCl₃ in S], ΔG_f^0[Li in B], ΔG_f^0[M in B] and ΔG_f^0[LiCl in S]. They are defined as the standard Gibbs energy changes associated with the formation of infinitely diluted species of them. In Eq. (3), ΔG_f^0[X, *liq*] represents the standard Gibbs energy change of formation of liquid states of component X.

$$\Delta G_f^0[\text{MCl}_3 \text{ in S}] = \Delta G_f^0[\text{MCl}_3, liq] + \Delta G^{ex}[\text{MCl}_3 \text{ in S}] \tag{3}$$

$$\Delta G_f^0[\text{Li in B}] = \Delta G_f^0[\text{Li}, liq] + \Delta G^{ex}[\text{Li in B}]$$

$$\Delta G_f^0[\text{LiCl in S}] = \Delta G_f^0[\text{LiCl}, liq] + \Delta G^{ex}[\text{LiCl in S}]$$

Substituting Eq. (3) to ΔG_{ext}^0 of Eq. (2) yields Eq. (4), which is the basic thermodynamic expression of the extraction reaction.

$$\log(D_M / D_{Li^3}) = -\frac{1}{2.3RT} \Delta G_f^0[\text{M in B}] \tag{4}$$

$$+\frac{1}{2.3RT}\left\{\Delta G_f^0[\text{MCl}_3, liq] + \Delta G^{ex}[\text{MCl}_3 \text{ in S}]\right\}$$

$$+\frac{3}{2.3RT}\left\{\Delta G_f^0[\text{Li}, liq] + \Delta G^{ex}[\text{Li in B}]\right\}$$

$$-\frac{3}{2.3RT}\left\{\Delta G_f^0[\text{LiCl}, liq] + \Delta G^{ex}[\text{LiCl in S}] + 2.3RT \log \gamma_{LiCl}\right\}$$

Results and discussion

The electromotive force measurement of Ho, Tb and La in liquid Bi

The electromotive force between M (solid) and M (in B) can be given by Eq. (5), where $\mu_{M(in B)}^{ex}$ is excess chemical potential of M in B and $\Delta\mu_M^{fusion}$ is the change of chemical potential over fusion of the metals M which equals the standard Gibbs energy change of fusion of M (ΔG_M^{fusion}).

$$\Delta E = \frac{1}{nF}\Delta\mu_M^{fusion} - \frac{RT}{nF}\ln x_{M(in B)} - \frac{1}{nF}\mu_{M(in B)}^{ex} \tag{5}$$

$\mu_{M(in B)}^{ex}$ is the excess function arising in mixing two liquids of M and B, but actually equals the excess Gibbs energy change of M in B ($\Delta G^{ex}[\text{M in B}]$). Eq. (5) yields Eqs. (6) and (7) where $f_{M(in B)}$ is the activity coefficient of M in B. By applying the observed ΔE and $\log x_{M(in B)}$ to Eq. (6), $\Delta G^{ex}[\text{M in Bi}]$ for Ho, Tb and La were calculated at 873 K and 1 073 K. The results are plotted in Figure 1 together with the $\Delta G^{ex}[\text{M in Bi}]$ of other lanthanides which were calculated by the reported activity coefficients with an adequate correction[6].

$$\Delta G^{ex}[\text{M in B}] = -nF\Delta E - \Delta G_M^{fusion} - RT\ln x_{M(in B)} \tag{6}$$

$$\Delta G^{ex}[\text{M in B}] = RT\ln f_{M(in B)} \tag{7}$$

In Figure 1, it was found that the $\Delta G^{ex}[\text{M in Bi}]$ shows close-to-linear dependence on the 2/3 power of the metallic molar volume of lanthanides ($V^{2/3}$). Through this linear relation, some $\Delta G^{ex}[\text{M in Bi}]$ for unreported lanthanides were estimated. A similar linear dependence can also be found in the case of liquid Zn. $\Delta G_f^0[\text{M in B}]$s which was obtained by correcting $\Delta G^{ex}[\text{M in Bi}]$ with ΔG_M^{fusion} are listed in Table 1. The observed close-to-linear dependence of ΔG_f^0 [M in B] on $V^{2/3}$ agrees with the Miedema's semi-empirical model for calculating the enthalpy change of solution of metal A into metal B (ΔH_{sol}[A in B]) [7,8].

Results of the liquid-liquid extraction experiment

Both in Zn and Bi systems, about ten to twenty pairs of $logD_M$ and $logD_{Li}$ for some trivalent lanthanides and some trivalent actinides (Pu, Np, Am and Cm) were measured at two different temperatures, 873K and 1 073K. The averaged $log(D_M/D_{Li}^3)$ values of elements were obtained as an intercept of the third-powered linear dependence of $logD_M$ on $logD_{Li}$ which were determined by applying the least square method of curve fitting to the sets of the $logD_M$ and $logD_{Li}$ data. The obtained $log(D_M/D_{Li}^3)$s of trivalent lanthanides and actinides are summarised in Table 2 [1,2].

Estimation and sytematic variation of ΔG^{ex} [MCl$_3$ in S]

By adapting experimentally determined $log(D_M/D_{Li}^3)$ and most probable values of ΔG_f^0 [M in B] to Eq. (5), with the use of other reported parameters [9-11], ΔG^{ex}[MCl$_3$ in S] for some lanthanides were calculated. The calculated values of ΔG^{ex}[MCl$_3$ in S] of lanthanides at two temperatures are plotted in Figures 2 and 3 as functions of the reciprocal of element's ionic radius. The ionic radii of the elements used are those of tri-positive cations under co-ordination number 6 [12]. In Figures 2 and 3, it is noteworthy that ΔG^{ex}[MCl$_3$ in S] obtained from two independent experimental systems (Zn and Bi) agree quite well, which suggests that this method of estimating ΔG^{ex}[MCl$_3$ in S] using distribution ratios is quite successful. The fact that ΔG^{ex}[MCl$_3$ in S] ranges from c.a. -14 to -90 kJ/mol suggests the presence of a quite strong chemical stabilisation of trivalent cations in the molten chloride phase [13].

It should be noted that ΔG^{ex}[MCl$_3$ in S] showed a roughly monotonic decrease on 1/R except for Pr and Lu, which showed certain discrepancies to the same direction from the lines. The linear tendency over the majority of the elements suggests that the stabilisation of lanthanide tri-positive cations presumably have an inverse dependence on the ionic size, which agrees with the expected coulombic co-ordination of chloride anions.

The value of ΔG^{ex}[MCl$_3$ in S] for some lanthanides gives certain discrepancies from those directly measured by electrochemical method [14]. In comparison with the ΔG^{ex}[MCl$_3$ in S], which were calculated from the results of standard potential measurement of M(0)-M(III) of some lanthanides [14], the currently estimated ΔG^{ex}[MCl$_3$ in S] is lower by about 25 kJ/mol. We have not yet identified the reason for this, but it may be attributed to the presence of a bias in the quoted values of the activity coefficients for Li or LiCl in this study, or to the reliability of the reference electrode of the literature [14].

Estimation of ΔG_f^0 [M in B] of actinides

Due to the experimental difficulty in handling massive actinide, the direct measurement of ΔG_f^0 [M in B] of actinides in liquid Bi or Zn has been limited. However, if we can estimate the ΔG^{ex}[MCl$_3$ in S] of trivalent cations of actinides in the molten chloride phase by applying the observed distribution ratios and estimated ΔG^{ex}[MCl$_3$ in S] to Eq. (4), we can calculate their ΔG_f^0 [M in B].

There is no concrete theoretical base to estimate $\Delta G^{ex}[MCl_3$ in S$]$ of trivalent actinides, but as a first order approximation, we can assume that actinide trivalent chlorides also satisfy the same dependence of $\Delta G^{ex}[MCl_3$ in S$]$ on the ionic radius as observed for lanthanides. According to some reported formal potentials of M(0)-M(III) for tri-positive U, Pu, Np, Nd and La in molten chloride [14], the estimated excess Gibbs energies of U, Pu, Nd and La show a rough linear dependence on 1/R, which agrees with the above assumption. However, Np is likely to show a singularity to this systematic [14], thus the application of the linear relation of lanthanide to Np is considered difficult.

By applying the ionic radii of Pu^{3+}, Am^{3+} and Cm^{3+} to the linear relation of $\Delta G^{ex}[MCl_3$ in S$]$, we estimated $\Delta G^{ex}[MCl_3$ in S$]$ for these three actinide trichlorides. Then, by applying these values to Eq. (4), we estimated $\Delta G_f^0[M$ in Bi or Zn$]$ of the actinides. Np, whose $\Delta G^{ex}[MCl_3$ in S$]$ was considered to be irregularly apart from the line of lanthanides, was omitted from this procedure. The results are shown in Figures 4 (Bi) and 5 (Zn) together with those of lanthanides as functions of $V^{2/3}$. It should be noted that, in Zn system at two different temperatures, Am, Cm and Pu seem to agree with the linear relation of lanthanides. On the other hand, in the Bi system, Am, Cm and Pu seem to be much lower than the line of lanthanides. These results suggests that the actinide metals gain more stabilisation in Bi solution than lanthanides do, and that they show different systematics on their metallic volume. This implies the presence of additional intermetallic bonding of Bi to actinides which is not found in the case of lanthanides, suggesting the presence of the contribution of 5f orbital with Bi. In contrast, in the Zn solution, actinides and lanthanides show close monotonic dependence on the metallic volume. This indicates that the intermetallic interaction of both lanthanide and actinides with Zn are simply controlled by their metallic volume, and there is no specific stabilisation due to 5f orbital.

Conclusions

We discussed the thermodynamic stabilities of actinides and lanthanides in liquid Bi and Zn in conjunction with those in molten chloride. By examining our own results of the EMF measurement and the literature data, we conclude that $\Delta G_f^0[M$ in Bi or Zn$]$ of lanthanides has a close-to-linear dependence on the 2/3 power of their metallic volumes. By applying the experimentally observed distribution ratios, we estimated $\Delta G^{ex}[MCl_3$ in S$]$ of lanthanides and found that they show linear dependence on the reciprocal of ionic radius. Using this systematic trend of the thermodynamic quantities, $\Delta G_f^0[M$ in Bi or Zn$]$ for Pu, Am and Cm were estimated, and deeper stabilisation of actinides in Bi was pointed out. The usability of the distribution data of pyrochemical liquid-liquid extraction for the estimation and evaluation of related thermodynamic quantities was proposed.

Acknowledgements

This study was supported by a Grant-in-Aid for Scientific Research provided by the Ministry of Education, Science, Sport and Culture of Japan. The authors are indebted to Mr. Sataro Nishikawa, Dr. Keizo Kawamoto, Dr. Jitsuya Takada, Mr. Hisao Kodaka and Mr. Kiyomi Miyata of Research Reactor Institute, Kyoto University for their technical support in the experimental activities. The authors also wish to thank Dr. Masaki Kurata for his suggestions and discussions concerning thermodynamic subjects.

REFERENCES

[1] H. Moriyama, H. Yamana, S. Nishikawa, Y. Miyashita, K. Moritani, T. Mitsugashira, *J. Nucl. Mater.*, 247 (1997), pp. 197-202.

[2] H. Moriyama, H. Yamana, S. Nishikawa, S. Shibata, N. Wakayama, Y. Miyashita, K. Moritani, T. Mitsugashira, *J. Alloys and Compounds*, 271-273 (1998), pp. 587-591.

[3] V.A. Lebedev, "Selectivity of Liquid Metal Electrodes in Molten Halides" (1993) (in Russian).

[4] V.I. Kober, V.A. Lebedev, I.F. Nichkov, S.P. Raspopin, *Russ. J. of Phys. Chem.*, 42 (3) (1968), 360.

[5] V.I. Kober, V.A. Lebedev, I.F. Nichkov, and S.P. Raspopin, *Russ. J. of Phys. Chem.*, 45 (3) (1971), 313.

[6] V.A. Lebedev, I.F. Nichkov, S.P. Raspopin, *Russ. J. of Phys. Chem.*, 42 (3) (1968), 363.

[7] F.R. deBoer, R. Boom, W.C. Mattens, A.R. Miedema, A.K. Niessen, "Cohesion in Metals – Transition Metal Alloys", North-Holland Physics Publishing, Amsterdam (1988).

[8] J.A. Alonso, N.H. March, "Electrons in Metals and Alloys", Academic Press, London (1989).

[9] I. Barin, O. Knacke, O. Kubaschewski, "Thermochemical Properties of Inorganic Substances", Springer, Berlin (1997).

[10] L.M. Ferris, J.C. Mailen, F.J. Smith, *J. Inorg. Nucl. Chem.*, 32 (1970), 2019.

[11] L. Lumsden, "Thermodynamics of Molten Salt Mixtures", Academic Press, London (1966).

[12] R.D. Shannon, *Acta Crystallogr.* A32 (1976), 751.

[13] H. Yamana, N. Wakayama, N. Souda, Hirotake Moriyama, *J. Nucl. Mater.*, 278 (2000), 37-47.

[14] Y. Sakamura, T. Hijikata, K. Kinoshita, T. Inoue, T.S. Storvick, C.L. Krueger, J.J. Roy, D.L. Grimmett, S.P. Fusselman, R.L. Gay, *J. Alloys and Compounds*, 271-273 (1998), 592.

[15] V.A. Lebedev, L.G. Babikov, S.K. Vavilov, I.F. Nichkov, S.P. Raspopin, O.V. Skiba, *Soviet Atom. Energ.*, 27(1) (1968), 748.

Table 1. $\Delta G°_f$[M in Bi] summarised (in kJ/mol)

	ΔG_f^0 [M in Bi] 873 K	ΔG_f^0 [M in Bi] 1 073 K
La	-197.62	-194.38
Ce	-184.10	-176.35
Pr	-179.75	-168.62
Nd	-180.04	-172.79
Pm		
Sm		
Eu		
Gd	-167.94	-163.24
Tb	-160.69*	-154.33*
Dy	-158.17*	-151.11*
Ho	-153.19*	-146.92*
Er	-143.79	-139.09
Tm	-147.42*	-140.68*
Yb	-150.81*	-144.73*
Lu	-143.48*	-136.17*

Estimated in this study

Table 2. Observed $\log(D_M/D_{Li}^3)$ of f-elements [1,2]

	Zn 873 K	Zn 1 073 K	Bi 873 K	Bi 1 073 K
La	10.879±0.065	8.903±0.061	6.605±0.063	5.328±0.111
Ce	11.197±0.119	9.226±0.072	6.713±0.099	5.668±0.111
Pr	11.578±0.030	9.558±0.056	6.911±0.067	5.664±0.114
Nd	10.935±0.165	8.920±0.243	6.648±0.250	5.608±0.111
Pm				
Sm	6.665±0.112	5.208±0.080	3.492±0.322	3.136±0.080
Eu	5.110±0.100	3.555±0.109	2.617±0.099	1.873±0.128
Gd	10.638±0.351	8.667±0.118	6.478±0.111	5.206±0.125
Tb	11.195±0.103	9.352±0.044	6.389±0.123	5.071±0.125
Dy		9.693±0.085		
Ho	11.010±0.175	9.298±0.091		
Er	10.829±0.197	9.386±0.111		
Tm	10.558±0.074	8.776±0.053	5.708±0.102	5.011±0.114
Yb	6.474±0.196	4.929±0.088	2.568±0.093	1.815±0.095
Lu	10.356±0.080	8.899±0.040		
Np	11.222±0.174	9.089±0.138	8.282±0.080	6.468±0.167
Pu	11.674±0.184	10.050±0.450	9.666±0.597	7.564±0.200
Am	12.353±0.205	10.309±0.159	9.485±0.202	8.056±0.139
Cm	11.984±0.208	10.040±0.170	9.113±0.150	7.690±0.109

Figure 1. Dependence of ΔG^{ex}[M in Bi] on $V^{2/3}$

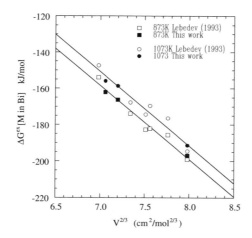

Figure 2. Estimated ΔG^{ex}[MCl$_3$ in S] at 873 K

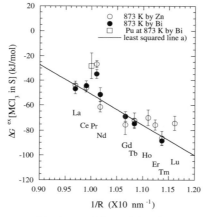

a) least squared line excluding Pr and Lu

Figure 3. Estimated ΔG^{ex}[MCl$_3$ in S] at 1 073 K

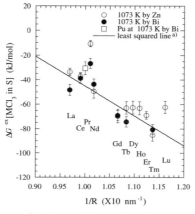

a) least squared line excluding Pr, Lu and Pu

Figure 4. Estimated ΔG_f^0 [M in Bi] of actinide

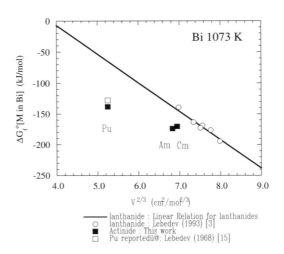

Figure 5. Estimated ΔG_f^0 [M in Zn] of actinide

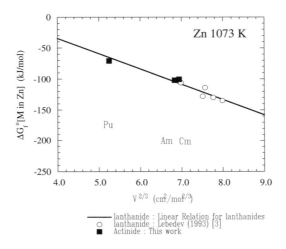

SESSION III

Part B: Results and Experiences from Process Experiments

Chairs: M. Gaune-Escard, T. Inoue

RESULTS OF DEMONSTRATION PROGRAMME OF ELECTROMETALLURGICAL PROCESSING OF BREEDER REACTOR FUEL

Gregory R. Choppin
Department of Chemistry
Florida State University
Tallahassee, FL 32306, USA

Abstract

Argonne National Laboratory (ANL) has developed an electrometallurgical technology (EMT) which is potentially applicable to a wide variety of spent nuclear fuel types. This process uses an alkaline metal chloride media and is directly applicable for metallic fuels. For other fuels (e.g. oxide fuels), it would be necessary to use head-end processes to convert the fuels to metals and to eliminate elements such as aluminium and carbon that, above certain concentrations, interfere with the electrometallurgical process in its present form. ANL has conducted a demonstration programme in which the development of the process was continued and a specified number of stored breeder reactor metallic fuel processed. In this demonstration programme 100 driver assemblies consisting of 410 kg of 60-75% highly enriched uranium and 25 blanket assemblies consisting of 1 200 kg of depleted uranium were treated. In the EMT process the chopped fuel elements are processed in an electrorefiner containing a KCl-LiCl molten salt eutectic. U, Pu, TRUs, the alkaline earth metals and rare earths are oxidised into the molten salt as anions. Subsequently, the U is reduced to the metal on a cathode. The salt containing the other cations and fission products are then mixed with zeolite and made into a ceramic waste form. The recovered uranium can be disposed in a repository or recycled for reuse.

Introduction

In processing of spent nuclear fuel two types of molten salt systems have been studied in the US and, on a small scale, used in the separations of the actinide and fission product elements. One process used a mixed fluoride salt fluid as coolant and homogeneous fuel and blanket system in the Molten Salt Reactor Experiment, and the second involved simple chloride eutectic salts as an ionic solvent in pyrochemical spent fuel reprocessing systems for highly irradiated metallic reactor fuels. However, at elevated temperatures, lithium metal in conjunction with lithium chloride reduces actinide oxides completely, and the resulting actinide metals can be processed by the same system as irradiated metallic reactor fuels.

An advantage of ionic molten inorganic salts and molten metals for use in fuel processing is that both phases are resistant to radiation damage effects. Another important advantage of pyrochemical treatment is the ability to chemically process "fresh" reactor fuels that have very high concentrations of fission products, with their associated high decay heat output and intense radiation emission. This decay heat, in fact, is useful as pyrochemical processing and is typically performed at process temperatures of 500 to 800°C. Extraction separations in the molten systems avoid the complications of hydrolysis and chemical instability of aqueous extraction chemistry. It is also useful that much of the needed thermodynamic data on the pyrochemical system is available, or is easily derived. An important aspect is the use of lithium chloride as part of the salt mixtures and lithium metal as a reducing agent. Magnesium chloride, copper chloride and cadmium chloride may be used, also, as selective oxidising agents.

In the Molten Salt Reactor Experiment application, a fluoride salt mixture of BeF_2, ^7LiF, ThF_4 and UF_4 was used as the molten fluid. Molten salt provided a fertile thorium blanket, a neutron multiplier, the fissile fuel and the reactor coolant, as well as much of the reprocessing solvent. It also served as a suitable solvent for the ^{233}Pa formed by neutron irradiation of the Th, and allowed its extraction from the salt phase into liquid bismuth by reduction with a controlled concentration of lithium metal. The fission products were removed from the flowing salt loop by processing a small stream diverted from the main coolant loop with a higher concentration of ^7Li metal. The purified salt was returned to the coolant stream after processing. The uranium was isolated as the volatile hexafluoride after fluorination with HF and fluorine. The iodine was removed as HI gas. The noble metals remained in their metallic form as a suspension in the salt.

The second type of process, developed at Los Alamos and Argonne National Laboratories, uses redox to separate and purify the uranium and plutonium from the other components in the spent fuel. In its simplest form, the electrochemical technique is based on the sequential anodic oxidation of the metals dissolved in a molten salt to cations, with transport of these cations through an ionic molten salt to the cathode where they are reduced to metal and deposited. The electromotive potential can be set within a narrow voltage window that permits either a single element to be deposited at the cathode, or, in the case of multiple cathodes, different elements to be deposited sequentially and selectively on separate cathodes. Impurity elements which require higher voltages in the molten salt or which undergo redox at lower voltages than the desired product remain in the anode pool, either in solution or as a sludge.

USDOE spent fuel

The USDOE has approximately 2 000 metric tonnes of spent nuclear fuel (SNF) stored in sites around the country. These SNFs are classified broadly as production fuels from the weapons programme, special fuels (e.g. from experimental reactors) and novel fuels. Not included in this inventory are the SNF from commercial reactors [1]. It has been proposed that a pyrochemical, electrometallurgical process under development at the Argonne National Laboratory for over 20 years could be used to process some fraction of the DOE SNF [2]. The electrometallurgical technology, EMT, developed by ANL is potentially applicable to a fairly wide variety of spent fuel types, if used in conjunction with appropriate head-end processes for non-metallic SNF. The purpose of these head-end processes would be to convert the fuels to metal and to free them from elements such as aluminium and carbon that, above certain concentrations, are incompatible with the electrometallurgical process as presently operated.

EMT has also been proposed by ANL as a process for use with actinide transmutation. A study of the use of EMT in such a transmutation process using reactors was conducted earlier by the National Research Council [3]. Modifications in the present technology are under study in transmutation systems which use accelerator irradiation [4].

In 1995, ANL initiated a demonstration programme of EMT as applied to stored driver and blanket fuel from the Experimental Breeder Reactor (EBR-II) activities. The USDOE requested that a committee established by the National Research Council provide oversight reviews during the period (four years) of the demonstration programme [5].

The electrometallurgical process

The electrometallurgical technique for treatment of DOE spent fuel, and in particular its application to the EBR-II Demonstration Project, evolved from earlier work on the Advanced Liquid Metal Reactor/Integral Fast Reactor (ALMR/IFR) [6]. This earlier work was directed to recycling IFR (and perhaps spent oxide fuels) into new IFR fuels, which would contain substantial quantities of uranium, plutonium, other actinides and long-lived fission products that would be burned in the IFR. A liquid cadmium cathode was to be used for separation of the bulk of the plutonium and the other TRUs from the bulk of the uranium, which was electrolytically deposited (separated) at a steel cathode. With the termination of the ALMR/IFR project, this process, with some modification, was proposed for the treatment of the EBR-II SNF. Use of the proposed process for oxide fuels would require a separate "front end", to convert the oxides into metal for use in the electrorefiner [6]. The proposed front end consists of reduction of the metal oxides into metal by Li, and electrochemical regeneration of metallic Li from Li_2O in molten LiCl. The process has also been considered as a possible technology for disposing of excess Pu from the US stockpile [7]. Figure 1 shows the process for the EBR-II Demonstration Program.

Electorefiners (Mark IV and V)

In the EBR-II SNF demonstration, 100 driver assemblies with a total of 410 kilograms of 60-75% highly enriched ^{235}U, and up to 25 blanket assemblies, consisting of 1 200 kilograms of depleted uranium, were to be processed. Chopped driver (or blanket) fuel rod elements are placed in a steel anode basket in an electrorefiner that contain molten KCl-LiCl eutectic at 500°C. The enriched uranium driver fuel is alloyed with ca 10 wt.% Zr; the cladding is stainless steel. The depleted uranium blanket fuel also has stainless steel cladding. The driver and the blanket fuel elements are sodium-bonded to the stainless steel cladding.

Two electrorefiners, Mark IV and V, have been used in the EBR-II Demonstration Project. The Mark IV (Figure 2) was used to electrorefine the driver elements from the EBR-II and has a molten Cd pool while the Mark V, used for the blanket elements, is Cd free. The Cd pool provides a corrosion resistant barrier to the mild steel vessel and serves as a neutron absorber to prevent criticality problems, which could result if the highly enriched uranium in the driver elements were to collect in the bottom of the vessel. The Cd pool in the Mark IV electrorefiner was not used as a cathode, and no Pu separation was performed.

The configuration of the Mark V differs markedly from the Mark IV, as it was designed to process much larger batches of material in the blanket elements. It is termed a "high throughput electrorefiner" (HTER) (Figure 3). However, the fundamental principles of the two electrorefiners are much the same. An oxidant, either $CdCl_2$, in the case of the Mark IV, or UCl_3 in the case of the Mark V, is added to the salt prior to electrolysis. The $CdCl_2$ oxidises some of the U (and other active metals) from the anode baskets. Upon passage of a constant electrolysis current between the anode baskets and the steel cathode, the U, Pu, TRUs, alkali, alkaline earth and rare earths metals are oxidised into the molten salt as U^{3+}, Pu^{3+}, TRU, alkali, alkaline earth and rare earth cations. The stainless steel from the cladding, most of the Zr and the noble metals remain in the anode baskets.

The U is eletrotransported from the chopped fuel (or blanket) elements in the anode basket to the cathode. The U^{3+} is reduced to the metal at the cathode and deposited in a reasonably pure state. The electrolysis is done under controlled current conditions so U^{3+} is selectively reduced at the cathode. To do this, a controlled amount of U^{3+} must be maintained in the melt. However, the build-up of sodium ion in the LiCl-KCl eutectic raises the melting point of the initial eutectic salt and, eventually, some of the salt must be removed and fresh KCl-LiCl added.

After a period of electrolysis, the U cathode is removed from the Mark IV electrorefiner, and the adhering salt volatilised in a vacuum furnace (the cathode processor) prior to reuse in the electrorefiner. The U is cast into an ingot in a high temperature furnace (the casting furnace). In the Mark V, the U deposits on a cathode from which it is continuously scraped into a collection basket beneath the electrolysis cell. Ultimately, the material remaining in the anode basket, consisting of the stainless steel hulls and any unoxidised material, the noble metals and some fission products, is also subjected to treatment to volatilise the adhering salts, and is then cast into an ingot to yield the metal waste form. The metal waste form should contain ca 15 wt.% Zr, which requires that material from the blanket processing must have Zr added in the casting furnace. At an appropriate time, the salt containing the Pu, TRUs, alkalies, alkaline earths and some fission products is removed from the refiner, mixed with zeolite, heated to adsorb the salt into the zeolite, mixed with glass and hot isostatically pressed into a glass bonded sodalite (GBS) the ceramic waste form.

Table 1 lists the free energies of formation per gm-equivalent of a number of the chlorides whose metals are important constituents of the spent nuclear fuel in the EBR-II. The elements are listed in three columns, proceeding from chlorides which are most stable to those least stable. On treatment of the SNF in the anode baskets, the metallic elements in the first two columns (the exception being Zr) are converted to chlorides and dissolve in the KCl-LiCl eutectic. The elements in the last column are sufficiently noble that they remain in the anode baskets in the metallic state. The electrolysis is carried out under a constant, controlled current. Limits are placed on both the anode and cathode potentials, such that Zr is essentially not anodised into the eutectic molten salt and only the U(III) is reduced at the cathode.

Were Zr to be oxidised and released into salt, it would be reduced at the cathode. The TRUs in the first column are oxidised into the molten salt but are not reduced at the cathode under the carefully controlled electrolysis conditions. The electrolysis currently does not use the liquid Cd as a cathode, but only as an anode to recover any U metal that is scraped from the steel mandrel and dissolves the liquid Cd pool.

During the repeatability phase of the Demonstration Project, the Mark IV electrorefiner was used to treat 12 driver assemblies at a rate of 24 kg of uranium per month. This rate exceeded the target criterion of 16 kg (~4 driver assemblies) per month for three months [8].

Each anode-cathode module (ACM) in the Mark V is capable of producing about 87-100 kg of uranium per month. The use of additional ACMs does not necessarily lead to a proportional increase in the monthly production rate of uranium. This is because only one ACM can be serviced at a time. Therefore, anode basket and product collector changeover must be carefully co-ordinated with the start of each run to avoid conflicts that might arise from the need to service two ACMs simultaneously. During the 30-day demonstration period, the Mark V ER processed the equivalent of 4.3 blanket assemblies/month. These blanket assemblies correspond to a total uranium deposit mass of 204.9 kg. During this time, 1.37 ACMs were active and the product collector rate was 212 g uranium/h/ACM. The average product collector can hold 13 kg of uranium product (uranium +20 wt.% salt), and three product collectors are required during the electrorefining of every 10 blanket elements.

Cathode processor

The purpose of the cathode processor is to remove entrained salt from the uranium electrodeposited and to consolidate the dendritic deposits. In the case of the driver fuel, depleted uranium must be added to the cathode processor to reduce the enrichment of the ER product to less than 50% (for security reasons). The cathode processor consists of an outer vessel, an induction-heated furnace assembly with coils, liner and insulation, a crucible assembly with a graphite process crucible and cover, radiation shield, condenser and a receiver crucible. There is a graphite liner in the induction furnace, and the induction coils are passively cooled and protected by a vapour barrier. Entrained salt and any cadmium present in the driver fuel are removed from the molten uranium ER product by vacuum distillation and deposited as a liquid in the receiver crucible. For convenience, the cathode processor is elevated above floor-level so that the crucible assembly can be bottom-loaded into the induction furnace. This position permits the process crucible to be loaded, emptied and cleaned without affecting the furnace assembly.

Casting furnace

The casting furnace provides a means to further reduce the enrichment of the driver fuel product resulting from the cathode processor, and to further consolidate the uranium product. The components of the casting furnace are similar to those of the cathode processor except that there is no condenser stage and associated receiver crucible to collect the distillate. Like the cathode processor, the casting furnace has an induction furnace and a graphite crucible, and can be evacuated. The furnace has two gas-tight flanges (top and bottom) that permit access to the casting crucible and other internal components. The crucible is normally loaded through the top flange using the same fixture which unloads the cathode processor.

Waste forms

Metal waste form

Following the fuel chopping and electrorefining operations, the cladding hulls are left in the anode basket, along with the noble metal fission products (Zr, Mo, Ru, Rh, Pd, etc.) and adhering salt. Actinides that remain within the cladding hulls are also present in the metal waste after electrorefining.

The uranium content is about 4% by weight. Zr metal is added to improve performance properties and to produce a lower melting point alloy. The target composition is stainless steel with an allowable range in Zr concentration of 5-20 wt.%. In order to distil the adhering processing salts, the material in the anode basket is placed in the cathode processor and heated to 1 100°C to distil the salt. The charge from the cathode processor is placed in an yttrium oxide crucible, melted at 1 600°C in the casting furnace in an Ar atmosphere, and either cooled in the crucible or cast into ingots. The actinides in the metal waste are primarily in this intermetallic phase. The resulting ingot, which is allowed to cool within the crucible, constitutes the metal waste form (MWF) [9].

Ceramic waste form

The ceramic waste form (CWF) has been developed to stabilise the active fission products (alkali, alkaline earths, and rare earths) and transuranic elements of the electrolyte [10]. The ceramic waste form is produced in a batch process by mixing and blending the waste salt, periodically removed from the electrorefiner, with zeolite 4A at 500°C to occlude the waste-loaded salt within the cages of the zeolite crystal lattice. Salt-loaded zeolite is mixed with a borosilicate glass and consolidated at high temperature (850-900°C) and pressure (14 500-25 000 psi) in a hot isostatic press (HIP) to make the final waste form. The minor component actinides and rare earths were found to form separate phases from sodalite and glass. The actinides occur as nano-size (colloidal) crystal inclusions associated with the glass or the glass/sodalite grain boundaries. Dissolution tests on the CWF over a six-month duration indicate that the CWF dissolves at a rate equal to or less than reference high level waste borosilicate glass. Also, tests showed that the mechanical and physical properties of the CWF are comparable to or better than borosilicate high-level waste glass.

Summary

The Demonstration Programme fulfilled the requirements established by the USDOE and ANL is planning the processing of the remaining EBR-II SNF. A decision on using the EMT process for other DOE SNF are under consideration.

REFERENCES

[1]   D.L. Fillmore and K.D. Bulmahn, "Characteristics of Department of Energy Spent Nuclear Fuel", in Proceedings of the Tropical Meeting on DOE Spent Nuclear Fuel, Salt Lake City, 13-16 Dec. 1994, American Nuclear Society, LaGrange Park, IL, p. 313ff. Unless otherwise noted, all data in this section are from this source.

[2]   "Proposal for Development of Electrometallurgical Technology for Treatment of DOE Spent Nuclear Fuel", Argonne National Laboratory, Idaho Falls, ID, 1995.

[3]   "Nuclear Wastes: Technologies for Separations and Transmutation", National Research Council, National Academy Press, Washington, D.C.

[4]   "A Roadmap for Developing Accelerator Transmutation of Waste (ATW) Technology, A Report to Congress", DOE/RW-0519, Department of Energy Office of Civilian Radioactive Waste Management, Washington, D.C., 1999.

[5]   "Electrometallurgical Techniques for DOE Spent Fuel Treatment: A Preliminary Assessment of the Promise of Continued R&D into an Electrometallurgical Approach for Treating DOE Spent Fuel" , National Academy Press, Washington, D.C., 1995.

[6]   J.P. Ackerman, T.R. Johnson, W.E. Miller, G. Burris, "Revised Flowsheet from Pyroprocessing Discharged IFR Fuel Materials,"ANL-IFR-79, Argonne National Laboratory, 1987.

[7]   "An Evaluation of the Electrometallurgical Approach for Treatment of Excess Weapons Plutonium", National Research Council, National Academy Press, Washington, D.C., 1996.

[8]   R.D. Mariani, D. Vaden, B.R. Westphal, D.V. Laug, S.S. Cunnigham, S.X. Li, T.A. Johnson, J.R. Krsul, M.J. Lambregts, "Process Description for Driver Fuel Treatment Operations", NT Technical Memorandum No. 111, Argonne National Laboratory, Argonne, IL, 1999.

[9]   "Metal Waste Form Handbook", D.P. Abraham, ed., ANL-NT-73 (Rev. 1), Chem. Tech. Div., Argonne National Laboratory, 1999.

[10]   "Preliminary Report on the Properties of Glass-Bonded Sodalite", Chem. Tech. Div., Argonne National Laboratory, 1999.

Figure 1. Diagram of the electrometallurgical process for the EBR-II SNF demonstration

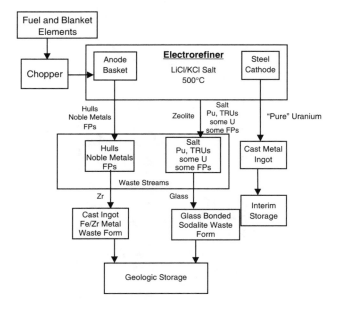

Figure 2. Schematic of the Mark IV electrorefiner

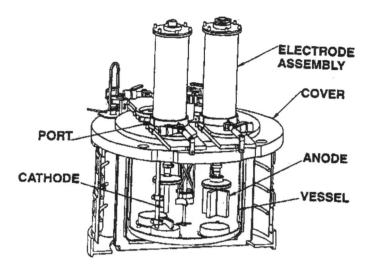

Figure 3. Anodic dissolution baskets in the Mark V electrorefiner

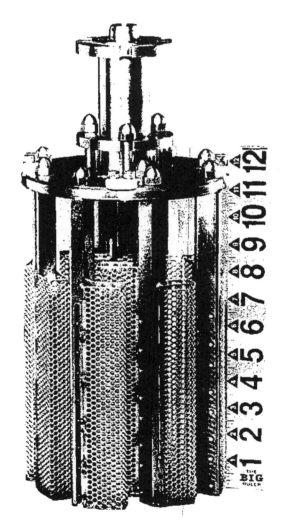

Table 1. Free energies of formation, -ΔG; of SNF chlorides (kcal/g-eq) at 500°C

I		II		III	
$BaCl_2$	87.9	$CmCl_3$	64.0	$CdCl_2$	32.3
$CsCl$	87.8	$PuCl_3$	62.4	$FeCl_2$	29.2
$RbCl$	87.0	$AmCl_3$	62.1	$NbCl_5$	26.7
KCl	86.7	$NpCl_3$	58.1	$MoCl_4$	16.8
$SrCl_2$	84.7	UCl_3	55.2	$TeCl_4$	11.0
$LiCl$	82.5	$ZrCl_4$	46.6	$RhCl_3$	10.0
$NaCl$	81.2			$PdCl_2$	9.0
$CaCl_2$	80.7			$RuCl_4$	6.0
$LaCl_3$	70.2				
$PrCl_3$	69.0				
$CeCl_3$	68.6				
$NdCl_3$	67.9				
YCl_3	65.1				

NITRIDE/PYROPROCESS FOR MA
TRANSMUTATION AND FUNDAMENTAL DATABASE

Toru Ogawa, Yasuo Arai
Japan Atomic Energy Research Institute
Tokai-mura, Naka-gun, Ibaraki-ken, 319-1195, Japan

Abstract

In the Japan Atomic Energy Research Institute, we study the double-strata fuel cycle concept. Each stratum of the fuel cycle may evolve independently: the power-reactor fuel cycle will be optimised for the safe and economical use of plutonium, and the MA burner system for the efficient reduction of long-lived hazards. The MA nitride fuel may be used and processed by a molten salt electrorefining technique in the second-stratum MA burner system. The possibility of pyroprocessing MA nitrides by an electrorefining method is studied. The UN, PuN and NpN have been electrolysed in molten salts to recover the respective metals. It proved to be fundamentally easy to recycle the expensive N-15. However, some complicating factors have also been identified. The R&D efforts are summarised and the technical issues are discussed from the basic point of view.

Introduction

The study of partitioning and transmutation (P&T) in Japan, which has occurred within the OMEGA project since 1988, is entering the second phase. The feasibility of minor actinides (MA: Np, Am and Cm) recycling with the pyrochemical separation is one of the key research subjects in this second phase. In the Japan Atomic Energy Research Institute, we study the double-strata fuel cycle concept (Figure 1) [1]. The MA are partitioned from the high-level wastes of the power-reactor fuel cycle (first stratum), and fed into the actinide burner cycle (second stratum), exiting only after transmuted to fission products. Each stratum of the fuel cycle may evolve independently: the power reactor fuel cycle will be optimised for the safe and economical use of plutonium, and the MA burner system for the efficient reduction of long-lived hazards. The MA nitride fuel may be used and processed by a molten salt electrorefining technique in the second-stratum MA burner system.

Since one has to deal with a significantly large decay heat and fast neutron emission in any MA burner cycle compared with the conventional fuel cycle, an innovative approach may be required in the fuel reprocessing. This is one of the reasons why the development of pyrochemical separation is important in the P&T study.

Actinide mononitrides are characterised by higher heavy-metal and smaller light-element densities compared with the oxides. Thus we may achieve improved neutron economy with the nitride fuels. This advantage can be utilised for more efficient burning/breeding. Another important advantage is the mutual mixability among the actinide mononitrides, in contrast to the metallic fuels where the mutual solubility between lighter actinides (U-Np) and heavier trivalent actinides (Am-Cm) is limited [2]. Thus, we may expect greater flexibility in designing the dedicated actinide-burning core. The isotope vector of the actinides strongly depends on the commercial spent fuel history as well as the recycle mode in a burner system. The fuel of a dedicated burner should accommodate a wide range of the combination and composition of actinides. The mutual mixability of actinide mononitrides has been demonstrated recently by the fabrication of (Cm,Pu)N [3], while Cm and Pu are almost non-mixable as solid metals.

The MA nitrides will be fabricated by a carbothermic synthesis from the MA salts coming from the first-stratum power reactor fuel cycle. One of the drawbacks of using nitride fuels is the cost of highly N-15 enriched nitrogen. The use of N-15 may be required, since C-14 produced by the neutron irradiation of N-14 is a long-term radiological concern. The possibility of applying the pyroprocess to the MA nitrides is being studied. The reference process flowsheet, which is currently being studied, is shown in Figure 2. In this reference flowsheet, the separation of actinides from fission products is achieved by molten salt electrorefining, which is similar to that for the metallic fuels [4]. In this way, development on the nitride/pyroprocess can be carried out in co-operation with that for metallic fuel.

Experiments for process development

Electrolysis of UN, PuN and NpN

The irradiated nitride fuels will be submitted for molten salt electrorefining to remove the fission products.

Following the demonstration of anodic dissolution of UN in the LiCl-KCl eutectic melt [5], the same techniques have been tried on PuN and NpN [6,7]. The temperature for electrolysis was about 773 K for these nitrides.

In the case of UN, we later identified the formation of ternary compound UNCl along with U^{3+} at the anode [8], which would adversely affect the efficiency of uranium recovery at the cathode. The U-free fuels would be used for the dedicated MA burner system, but this problem has to be solved for the potential application of nitride/pyroprocess for the FBR fuel cycle. It should be mentioned, however, that UNCl is hardly soluble in the LiCl-KCl eutectic melt, and recovered for the later thermal decomposition to produce UN.

The formation of ternary compound MNCl is considered unlikely for the transuranium elements as elucidated for Pu in Figure 3. Actually, the concentration of Pu in the molten salt remained almost constant in the electrolysis runs of PuN. In addition, at the constant-potential electrolysis runs, the current increased with time due to the growth of cathode deposit. Hence, we did not notice any effect which might accompany the formation of insoluble nitride chloride. In these runs the anode was a perforated tungsten cage containing a sintered PuN specimen, and the cathode was a solid molybdenum rod. The recovered deposit was a mixture of Pu metal and salts. The results were similar for NpN.

Nitrogen release at dissolution of metal nitrides

The cost accompanying the use of N-15 may become prohibitive, unless one devises means to recover and recycle it in the fuel cycle. While N-15 may become diluted with natural nitrogen in ordinary aqueous processes, it may be readily recovered in a dry process such as the molten salt electrorefining.

This point has been demonstrated by dissolving NdN and DyN, which are regarded as surrogates of transplutonium nitrides, by oxidising them with $CdCl_2$ in the LiCl-KCl eutectic melt [9]. Nitrogen contained in DyN was released for 97-99%, while that in NdN slightly less than these values. In these tests, NdN powder was commercially obtained, but DyN was prepared in JAERI to reduce the oxygen content to a level of 0.5 wt.%.

The difference in behaviour between NdN and DyN was considered due to the larger amount of oxygen impurities in NdN. A Nd-concentrated sediment was found at the bottom of the salts. It is likely that some nitrogen exists in the bottom sediment along with Nd, O and Cl. It was confirmed by the high-temperature UV-vis spectrophotometry that an impure NdN reacts with $NdCl_3$ in the LiCl-KCl eutectic melts as inferred from Figure 4 [10].

Nitridation in liquid cadmium

The actinide metals recovered by molten salt electrolysis will be converted to nitrides. This will be done through nitridation in liquid metals such as cadmium. The existing thermodynamic data on the nitrides, chlorides and liquid-Cd alloys of actinide and rare-earth elements indicate:

- The free energy of formation of MN is rather close to each other among these elements.

- Those of MCl_3 are significantly more negative for the rare earths compared with actinides.

- The activity coefficients, γ, of metals in liquid Cd alloys are more negative for the rare earths compared with actinides. Among the rare earths, γ becomes larger with decreasing ionic size.

Nitrogen was fed into liquid Cd-2 wt.%U-1 wt.%Gd-1 wt.%Ce alloy at 773-873 K. The alloy was contained in either graphite or molybdenum container [11]. Uranium was preferentially converted to U_2N_3. A little fraction of Gd precipitated as either (U,Gd)N or GdN, but the lanthanides mostly remained dissolved in Cd. Particularly, Ce was the last to react with nitrogen. The behaviour follows theoretical expectation based on the above thermodynamic characteristics for MN and Cd-M alloys. The molybdenum container was found more suitable for the process. With the graphite container, the products were contaminated with carbon.

Ongoing and future experimental programmes

So far the experiments on molten salt electrolysis have employed solid cathode. The liquid metal cathode may be used to recover the transuranium elements. The electrolysis of Pu with liquid Cd electrodes is being done as a joint study with the Central Research Institute of Electric Power Industry (CRIEPI). The study will be extended to the electrolysis of PuN and NpN.

The nitrides of Np, Pu, Am and Cm are routinely prepared for various basic experiments, although AmN and CmN are prepared only in a 10 mg scale due to the necessary radiation shielding and the availability of these transplutonium elements. During FY2000, the capability to analyse AmN and CmN for carbon and oxygen impurities will be obtained.

It will become possible to handle gram-quantities of Am in FY2002. A shielded facility dedicated to the study of the high-temperature chemistry of Am and Cm will be constructed in the Tokai Establishment, JAERI, in FY2000-2001 as a part of joint programme with the Japan Atomic Power Company.

Thermodynamic datafile

It has become fairly routine to predict high-temperature processes through thermodynamic analysis with a Gibbs free energy minimiser. In JAERI, the emphasis is to prepare the thermodynamic datafile for nitride/pyroprocess, which is consistent with the experimental data from electrochemical measurements.

The whole data should be summarised in an input file for the free energy minimiser *ChemSage*: "LiCl-KCl-MCl$_x$-CdCl$_2$-MN-Cd-M-N-C-O" (M: actinides and rare earths). The data in this file should be compared with the behaviour of various reference systems:

- "LiCl-KCl-Li-O-N" for oxygen and nitrogen behaviour.

- "LiCl-KCl-MCl$_3$-MCl$_2$-M" for $M^{2+}/(M^{3+}+M)$ equilibria.

- "LiCl-KCl-MCl$_3$-MOCl-M$_2$O$_3$" for the interference of impurity oxygen.

- "U-N-Cl" for the stability of MNCl.

- "M-N-C-O" for the stability of oxygen in the nitrides.

- Etc.

Interest in these subsystems is in common with the other pyroprocesses on oxide and metallic fuels. Figure 5 shows the calculated E-p(O^{2-}) diagram of Pu in LiCl-KCl eutectic melts, for example. Such prediction should fail if there is a significant association between metal ions and oxide ion in the molten salts. Concerted efforts among those who are interested in the pyroprocessing of the nuclear fuel should be due, regardless of the difference in the assumed scenarios.

REFERENCES

[1]   T. Mukaiyama, *et al.*, "Partitioning and Transmutation Program "OMEGA" at JAERI," Proc. Int. Conf. GLOBAL'95, 11-14 September 1995, p. 110, American Nuclear Society Topical Meeting.

[2]   T. Ogawa, M. Akabori, F. Kobayashi and R.G. Haire, "Thermochemical Modeling of Actinide Alloys Related to Advanced Fuel Cycles, *J. Nucl. Mater.* 247 (1997), 215-221.

[3]   M. Takano, *et al.*, paper to be presented at the AESJ Spring Meeting, 28-30 March 2000.

[4]   Y.I. Chang and C.E. Till, "Actinide Recycle Potential in the Integrated Fast Reactor Fuel Cycle", pp. 129-137 in LMR: A Decade of LMR Progress and Promise, American Nuclear Society, Inc., La Grange Park, IL, 1990.

[5]   F. Kobayashi, *et al.*, "Anodic Dissolution of Uranium Mononitride in Lithium Chloride-Potassium Chloride Eutectic Melt," *J. Am. Ceram. Soc.*, 78, 2279 (1995).

[6]   O. Shirai, *et al.*, "Electrolysis of Plutonium Nitride in LiCl-KCl Eutectic Melts", *J. Nucl. Mater.* 277 (2000), 226-230.

[7]   Y. Arai, *et al.*, "Experimental Research on Nitride Fuel Cycle in JAERI", Proceedings of the International Conference on Future Nuclear Systems, 29 August-3 September 1999, Jackson Hole, Wyoming.

[8]   F. Kobayashi, *et al.*, "Stability of UNCl in LiCl-KCl Eutectic Melt", *J. Alloys and Compounds*, 271-273 (1998), 374-377.

[9]   F. Kobayashi, *et al.*, "Dissolution of Metal Nitrides in LiCl-KCl Eutectic Melts", Proceedings of the International Conference on Future Nuclear Systems, 29 August-3 September 1999, Jackson Hole, Wyoming.

[10]   M. Matsumiya and T. Ogawa, *unpublished study*.

[11]   M. Akabori, A. Itoh and T. Ogawa, "Nitridation of Uranium and Rare-Earth Metals in Liquid Cd", *J. Nucl. Mater.*, 248 (1997), 338-342.

Figure 1. Double-strata fuel cycle concept

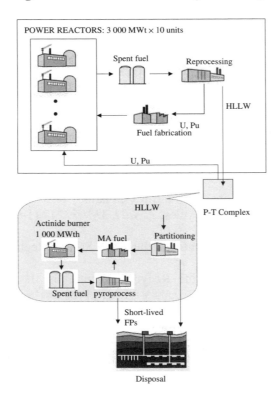

Figure 2. Reference flowsheet for processing nitride fuels

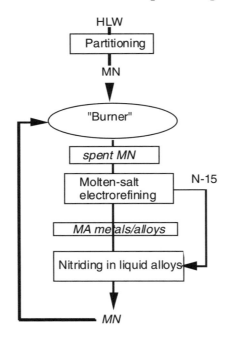

Figure 3. Stability diagram of U-N-Cl system

Formation of PuNCl is less likely due to the greater stability of PuCl₃
and lesser stability of Pu⁴⁺ compared with the system with uranium

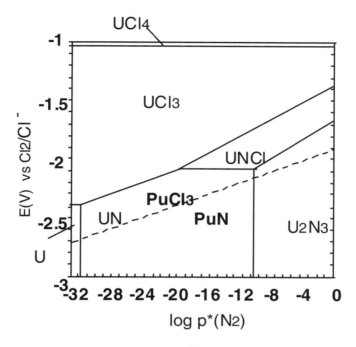

Figure 4. High-temperature UV-vis spectra of Nd³⁺ in LiCl-KCl-NdCl₃ melt with NdN powder

NdN having the overall composition $NdN_{0.912}O_{0.202}C_{0.007}$ reduces
absorption peaks of Nd³⁺ presumably due to the oxygen impurity

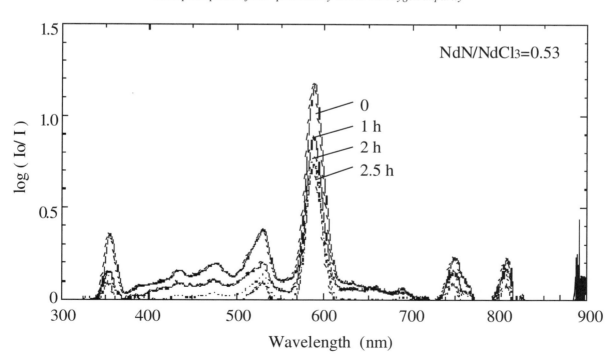

Figure 5. Predicted E-p(O^{2-}) diagram for Pu in LiCl-KCl eutectic melt at 773 K

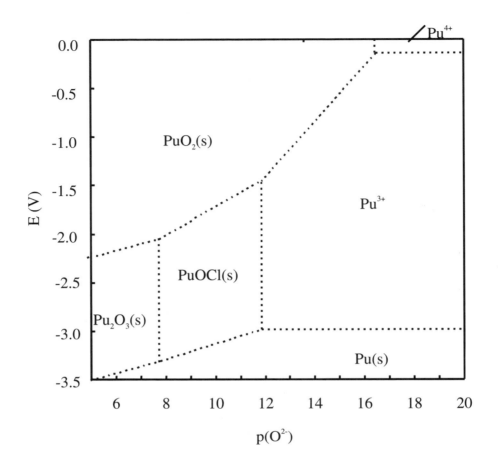

PYROMETALLURGICAL REDUCTION OF UNIRRADIATED TRU OXIDES BY LITHIUM IN A LITHIUM CHLORIDE MEDIUM

Tsuyoshi Usami, Masaki Kurata and Tadashi Inoue
Nuclear Fuel Cycle Department
Central Research Institute of Electric Power Industry
Japan

Jon Jenkins, Howard Sims, Steve Beetham and Denzil Brown
AEA Technology Nuclear Science
United Kingdom

Abstract

The Central Research Institute of Electric Power Industry (CRIEPI) is investigating Li reduction of oxide fuel because it is a candidate process for part of a PWR/metal FBR fuel cycle. One of the components of the investigation is a series of reduction experiments on unirradiated TRU oxides, which is being carried out under contract by AEA Technology.

This paper describes the facility built for the experiments and the following measurements that are used to determine the extent of oxide reduction to metal:

- The concentrations of Li_2O and TRU in samples of salt taken during an experiment.

- The quantity of TRU metal in samples of the final product.

The techniques required for these measurements have been successfully calibrated and demonstrated in pilot tests with UO_2. They are used in experiments with PuO_2, mixed UO_2 and PuO_2, AmO_2, NpO_2, and oxide mixtures of actinides and representative non-active fission products.

Introduction

Lithium has several advantages over calcium as a pyrochemical reductant for oxide fuel:

- It does not require ceramic containment material.

- There is a possibility that it will reduce actinide oxides but not rare earth oxides.

- It requires a lower temperature process.

Lithium, as a reductant, is being extensively investigated at Argonne National Laboratory (ANL) with uranium dioxide (UO_2) pellets in batches of up to 30 kg [1]. The reduction is carried out at 650°C in a molten LiCl medium. Lithium oxide (Li_2O) is produced by the reduction reaction and dissolves in the molten salt. Lithium metal is regenerated from the $LiCl/Li_2O$ product by electrowinning so that LiCl and lithium can be recycled during the reduction process.

However, thermodynamic assessment indicates that the reduction of PuO_2 and AmO_2 might not be complete. For example, as the data in Table 1 show, the Gibbs free energies of formation of PuO and the sesquioxide forms of Pu and Am are very close to that of Li_2O when Li_2O concentration is relatively high. Hence only partial reduction of PuO_2 and AmO_2, to Pu_2O_3/PuO and Am_2O_3 respectively, might occur unless Li_2O concentration is maintained at a sufficiently low level. Preliminary work at ANL with small quantities of synthetic LWR fuel appears to support this assessment [2].

CRIEPI is developing the Li reduction process, as an option for reducing LWR fuel prior to its purification by electrorefining and feeding into a metal FBR fuel cycle [3]. CRIEPI has contracted AEA Technology to construct a pyrometallurgical apparatus in its radiochemical facility at Harwell, to investigate the reduction of unirradiated actinide oxides with the participation of CRIEPI staff. This paper describes the following aspects of the work at Harwell:

- The facility and experimental reduction procedure.

- Measurements of Li_2O solubility in LiCl, to calibrate the system against published ANL data.

- Experimental techniques that are used to determine the extent of oxide reduction to metal.

- An outline of the experimental programme being carried out for CRIEPI.

The experimental techniques are illustrated with data from experiments with UO_2 and PuO_2.

Description of facility and experimental reduction procedure

The facility comprises two high-integrity glove boxes:

- A "dry box" containing the pyrochemical reduction furnace, which has an argon atmosphere with less than about 10 ppm oxygen and 25 ppm water vapour.

- A "wet box" where samples from the dry box are prepared for analysis, and in which oxygen levels are also kept very low.

The pyrochemical furnace has three individually controlled resistance heating coils and is located in a water-cooled stainless steel well attached to the underside of the glove box. A separate, high purity argon supply is fed to the furnace, which can accommodate a crucible up to 60 mm diameter by 152 mm long. However, for the experiments carried out to date the crucibles are 20 mm diameter by 152 mm long and used to reduce approximately 1 g to 2 g of powdered TRU oxide, or a single mixed actinide pellet weighing about 10 g.

A scheme of the general experimental arrangement for reduction is shown in Figure 1. Powdered TRU oxide is mixed with LiCl in a tungsten crucible at 650°C and Li metal added in small quantities at pre-determined time intervals. For some experiments, Li_2O was added to the initial reaction mixture to determine the effect of Li_2O concentration on the reduction process. Temperature is monitored by a thermocouple attached to the outside of the crucible and used to control the furnace temperature.

The arrangement for mixed actinide pellets is similar to that for powders, except that the pellet is held in a tantalum wire basket so that it can be suspended in the salt and recovered straightforwardly after the experiment.

Three devices are used for sampling the molten salt during an experiment:

- A stainless steel rod, also used to measure salt levels in the crucible.

- A stainless steel dip tube.

- An Inconel filter tube.

Level measurements are made by rapidly immersing a stainless steel rod into the melt through the central furnace penetration. Salt freezes upon the metal surface on withdrawal of the rod, and gives an indication of the level and state of the molten salt within the crucible. For example, pure LiCl is colourless and transparent, whilst LiCl containing Li_2O has a "whitish" appearance which darkens with increasing oxide content. Level samples are mechanically removed from the steel rod for analysis.

A dip tube is an open stainless steel tube which is used when a larger salt sample is required for analysis. A filter tube is similar but has a mesh of Inconel swaged onto one end. Filter tube samples are taken when there is a requirement to obtain salt samples that have no entrained particulates, for example, during the TRU and Li_2O solubility measurements.

On completion of a reduction experiment, the crucible is cooled and removed from the apparatus. The assembly is visually examined and weighed and, for most experiments, broken in a controlled way to isolate the contents for inspection and chemical analysis.

Lithium oxide solubility data

The solubility data were obtained in two separate experiments. The Li_2O was produced by calcining lithium carbonate in air at 1 000°C in a platinum crucible, and pure LiCl was sourced in one experiment from Fischer and in the other from Aldrich APL.

A mixture of 15.1 wt.% Li_2O in LiCl was prepared in an Inconel crucible. The salt mixture was heated to about 770°C for 20 hours, prior to cooling to 750°C, to ensure full dissolution had been achieved. Samples of salt were taken with filter tubes, which were suspended above the melt for some time to allow them to equilibrate thermally before immersion in the salt. The salt was dissolved in water and titrated with dilute hydrochloric acid to determine Li_2O concentration. Samples were taken in the following order of temperature; 750 (second experiment only), 700, 650 and 600°C. An Inconel stirrer was used to prevent the salt mixture from super-cooling during the temperature changes.

Results from the solubility tests are shown in Table 2. The agreement between the two sets of data is good, although the solubility at 700°C was about 5% higher for the second set. The Li_2O solubility, averaged over the two sets of data and with their standard deviations, varied from 7.53 wt.% ± 0.46 wt.% at 600°C to 10.21 wt.% ± 0.48 wt.% at 700°C, with a single value of 11.39 wt.% at 750°C.

The measurements compare well with those published by Johnson and co-workers [2] at Argonne National Laboratory. In particular, for the preferred operating temperature of 650°C, the solubility was 8.77 wt.% ± 0.31 wt.% compared with 8.70 wt.% measured at ANL.

Determination of extent of reduction

On-line analysis of Li_2O concentration

Samples of salt are taken from the melt during an experiment, weighed and dissolved in water. Since a small quantity of Li metal is likely to be present, the dissolution step is carried out in a gas burette, as shown in Figure 2. The hydrogen evolved is measured with the burette and the mass of Li metal determined from the following reaction:

$$2Li + 2H^+ \rightarrow H_2 + 2Li^+ \tag{1}$$

The product solution is then titrated with 0.1 M HCl to determine total Li. The estimated Li metal content is subtracted from the total Li measurement to determine the Li_2O content of the sample.

Figure 3 gives a typical plot of Li_2O concentration against the cumulative mass of Li metal added to the melt, for the case of UO_2 initially in pure LiCl. It can be seen that Li_2O concentration increased linearly and reached a maximum of about 2.5 wt.%.

The extent of reduction is then deduced from the calculated mass of Li_2O and the reduction stoichiomentry, for example:

$$UO_2 + 4Li \rightarrow U + 2Li_2O \tag{2}$$

Figure 4 gives the calculated UO_2 reduction corresponding to the Li_2O concentration data in Figure 3, and shows that the final Li_2O concentration of 2.5 wt.% corresponded to approximately 100% theoretical UO_2 reduction.

Analysis of TRU concentration in salt

The product solutions from titration of the salt with HCl are generally analysed by gamma spectrometry to determine TRU content. Titration produces clear product solutions and, therefore, all of the TRU in the salt appears to be soluble in the aqueous phase.

In the case of experiments with a single TRU species the following γ-ray energies are detected:

- UO$_2$ – 186 keV γ-ray (53% abundant) from the decay of ^{235}U.

- PuO$_2$ – 129 keV γ-ray (0.0062% abundant) from ^{239}Pu, with the 148 keV γ-ray (0.000185% abundant) from ^{241}Pu as a secondary check.

- AmO$_2$ – 59 keV γ-ray (36% abundant).

- NpO$_2$ – 29 keV γ-ray (14% abundant).

Tests with plutonium are complicated by the presence of ^{241}Am, formed by decay of ^{241}Pu, which leads to high detector dead-times. For experiments with mixed uranium and plutonium oxides, the ^{235}U γ-ray is swamped by the Pu and Am emissions and therefore U is analysed by inductively-coupled plasma/optical emission spectrometry (ICP-OES).

Figure 5 shows how the total Pu in the salt varied with Li metal added for an experiment with PuO$_2$. It can be seen that the total Pu was constant, at approximately 2 mg, throughout most of the experiment. When 200 mg of Li had been added, and plutonium reduction was almost complete, the Pu content of the salt decreased to < 0.1 mg. The Pu metal product separates from the salt as described in the next section.

Analysis of final product

The broken crucible and its contents are generally divided into the following three categories for sampling for analysis.

1. The salt at the base of the crucible, which contains finer particles of TRU oxide and/or metal.

2. Fragments of tungsten from the base of the crucible. These are coated with either:

    – A layer of compacted particles, for example, in the case of U.

    – A film of metal in the case of Pu, which has a melting point a little lower than the experimental temperature of 650°C.

3. Fragments of tungsten from the upper regions of the crucible. Liquid Pu wets tungsten and therefore forms a thin film up the sides of the crucible.

Each type of sample is washed with methanol in the gas burette to remove LiCl and to measure and remove lithium metal. The residual sample is then dried, weighed and contacted with aqueous hydrobromic acid in the gas burette. The quantity of TRU metal is calculated from the volume of hydrogen evolved, for example, according to the following idealised reaction stoichiometries:

$$U + 4H^+ \rightarrow 2H_2 + U^{4+} \tag{3}$$

and:

$$Pu + 3H^+ \rightarrow 1.5H_2 + Pu^{3+} \tag{4}$$

The gas burette was calibrated with samples of U and Pu metal and the above stoichiometry confirmed, although in the case of U a small correction factor is required.

The product solution from the gas burette is analysed by gamma spectrometry in order to measure the total TRU content of a sample, which can then be compared with the metal calculated from the gas burette to determine the percentage reduction for that sample.

Typical analytical results for a UO_2 reduction experiment are given in Table 3, which lists the weight of washed samples, their metal content derived from the volume of hydrogen measured by the gas burette, and the total uranium calculated from the activity measured by γ-spectroscopy of the product solution. It can be seen that there is close agreement between sample weight and metal content. Figure 6 compares the masses of uranium determined by the two analytical methods. It can be seen that there is very good agreement between the data: the points lie either side of the line of equality and only one point is slightly outside the $\pm10\%$ error bars. Overall, the data demonstrate that the analytical methods are satisfactory and reduction of oxide to metal was essentially complete.

A summary of the uranium analyses for this experiment is given in Table 4, which shows that:

- The layer of compacted particles at the bottom of the crucible contained 90% of the U solids recovered from the crucible and the average of the gas burette analyses indicated that 99% of the solids were metal.

- The salt close to the compacted particles contained 9% of the U solids recovered from the crucible and the average of the gas burette measurements indicated that 97% of these solids washed from the salt was U metal.

- The uranium adhering to the crucible fragments was only about 1% of the total U solids recovered from the crucible.

In the above case, an overall mass balance shows that 93% of the uranium added at the beginning of the experiment was accounted for by analysis.

Experimental programme

Reduction of the following unirradiated materials are being investigated:

- UO_2 powder.

- PuO_2 powder.

- Mixed UO_2 and PuO_2 (MOX) in pellet form.

- AmO_2 powder in isolation (as opposed to grown into PuO_2).

- NpO_2 powder.

- Pellets containing all of the above actinides, with Cm, representative rare earth, noble metal, alkaline and alkaline earth metal oxides.

The experiments consume about 1-2 g of powdered TRU or a 10 g pellet of mixed TRU. The effect of Li_2O concentration on reduction is being studied in most of the above cases.

Conclusions

- A series of experiments to reduce unirradiated TRU oxides with Li metal in a LiCl medium is being undertaken for CRIEPI by AEA Technology.

- The experiments are carried out at 650°C and extent of reduction is determined by measuring Li_2O and TRU concentrations in the salt during the experiment, and analysing the final product for its metal content.

- The system has been calibrated by obtaining Li_2O solubility data in close agreement with ANL data, and by demonstrating high reduction of UO_2.

- The facility is being used to determine the reduction of the major and minor actinides, in isolation and in combination, and with representative non-active fission products. The dependence of reduction on Li_2O concentration is being investigated.

REFERENCES

[1]   E.J. Karell, K.V. Gourishankar, L.S. Chow and R.E. Everhart, "Electrometallurgical Treatment of Oxide Spent Fuels", GLOBAL'99, International Conference on Future Nuclear Systems, 29 August-3 September 1999, Jackson Hole, Wyoming, USA.

[2]   G.K. Johnson, R.D. Pierce, D.S. Poa and C.C. McPheeters "Pyrochemical Recovery of Actinide Elements from Spent Light Water Reactor Fuel", p. 199 of "Actinide Processing: Methods and Materials", B. Mischra, ed., published by the Minerals, Metals and Materials Society, 1994.

[3]   T. Inoue and H. Tanaka, "Recycling of Actinides Produced in LWR and FBR Fuel Cycles by Applying Pyrometallurgical Process", GLOBAL'99, International Conference on Future Nuclear Systems, 29 August-3 September 1999, Jackson Hole, Wyoming, USA.

[4]   K.V. Gourishankar and E.J. Karell, "Application of Lithium in Molten Salt Reduction Processes", Light Metals 1999, C. Edward Eckert, ed., published by the Minerals, Metals and Materials Society, 1999.

Table 1. Approximate Gibbs free energies of oxide formation at 650°C

From data compiled in References [3] and [4]

Species	ΔG_T^0 (kJ/mole of O_2)
CmO_2	-740
AmO_2	-770
NpO_2	-870
PuO_2	-880
UO_2	-930
Li_2O (high concentration*)	-955
PuO	-960
Pu_2O_3	-970
Am_2O_3	-980
Li_2O (low concentration*)	-980

* Note: Data from Reference [4] are for varying Li_2O activity coefficient.

Table 2. Li₂O solubility test results

Temperature (°C)	Set	Li_2O concentration (wt.%)
750°C	2	11.39
700°C	1	9.91
	1	10.14
	1	10.14
	1	10.06
	2	10.54
	2	10.48
	Mean, σ	10.21±0.48
650°C	1	8.68
	1	8.90
	1	8.64
	2	8.80
	2	8.83
	Mean, σ	8.77±0.10
600°C	1	7.18
	1	7.70
	2	7.66
	2	7.59
	Mean, σ	7.53±0.46

Table 3. Comparison of analytical results for a UO₂ reduction experiment

Sample	Sample mass (g)	Mass of U metal from gas burette (g)	Mass of U by gamma activity (g)	% reduction
Compacted particles at bottom of crucible	0.072	0.071	0.069	97.2
	0.088	0.086	0.080	100*
	0.106	0.100	0.109	91.7
	0.08	0.082	0.079	100*
	0.06	0.06	0.057	100*
Salt containing fine particles, at bottom of crucible	0.118	0.12	0.113	100*
	0.043	0.039	0.037	94.9
Tungsten fragments from upper regions of crucible	–	0.0165	0.016	–

* Note: Shown as 100%, although division of the U metal measured with the gas burette by the total U is actually > 100% due to experimental error.

Table 4. Summary of analytical results for a UO₂ reduction experiment

Material	Mass of associated U solids (g)	Proportion of total U solids recovered (%)	Number of samples analysed	Mass of U metal by gas burette (g)	Proportion of U metal in mass of U solids
Compacted particles at bottom of crucible	1.72	90	5	1.70	99
Salt containing fine particles, at bottom of crucible	0.17(5)*	9	3	0.17	97
Tungsten fragments from upper region of crucible	0.02	1	1 (all of fragments)	0.02	≈100

*Note: These solids were washed from 2.21 g of salt.

Figure 1. Schematic diagram of pyrochemistry apparatus (not to scale)

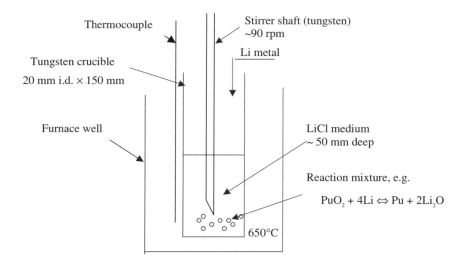

Figure 2. Gas burette apparatus for TRU metal determination (not to scale)

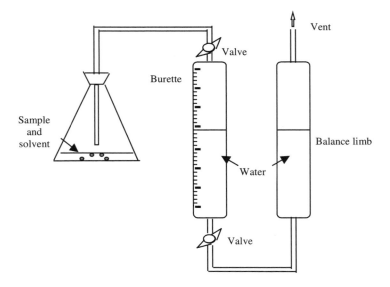

Figure 3. Li₂O concentration in salt as a function of Li added, for reduction of UO₂ particles

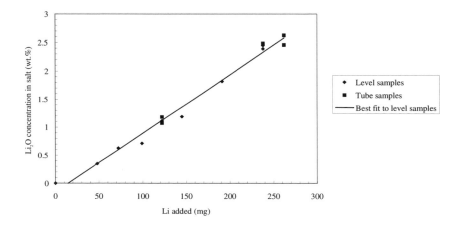

Figure 4. Theoretical UO₂ reduced as a function of Li added, for reduction of UO₂ particles

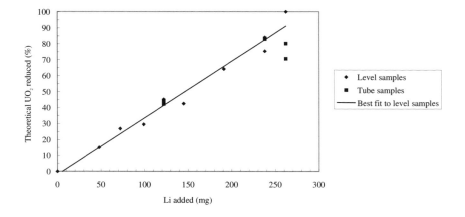

Figure 5. Variation of Pu concentration in salt as a function of Li added, for reduction of PuO₂ particles

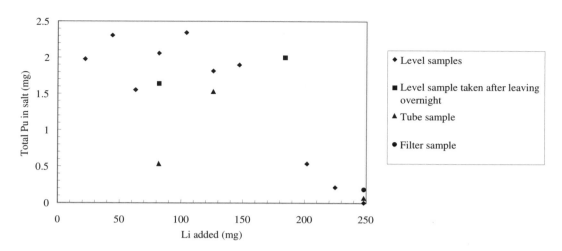

Figure 6. Correlation of sample analysis for a UO₂ reduction experiment

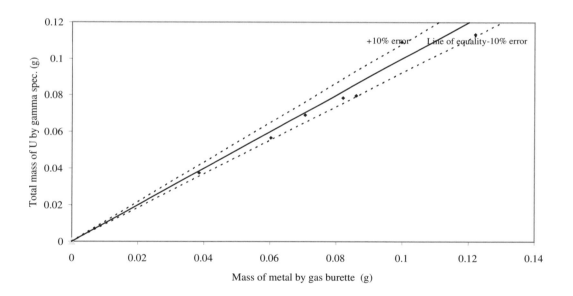

SMALL-SCALE DEMONSTARATION OF PYROMETALLURGICAL PROCESSING FOR METAL FUEL AND HLLW

Tadafumi Koyama*, Kensuke Kinoshita, Tadashi Inoue
Central Research Institute of Electric Power Industry (CRIEPI)
2-11-1 Iwado-kita, Komae, Tokyo 201-0004, Japan
*koyama@criepi.denken.or.jp / koyama@itu.fzk.de

Michel Ougier, Jean-Paul Glatz, Lothar Koch
European Commission Joint Research Centre
Institute for Transuranium Elements (JRC-ITU)
Postfach 2340, D-76125 Karlsruhe, Germany

Abstract

CRIEPI and JRC-ITU have started a new joint study on pyrometallurgical processing. The objective of this study is to demonstrate the capability of this type of process to separate actinide elements from spent fuel and HLLW. In the first phase, the following two different series of experiments will be carried out:

- Pyroreprocessing of unirradiated metal alloy fuel (U-Pu-Zr or U-Pu-MA-RE-Zr) by molten salt electrorefining and molten salt/liquid metal extraction.

- Pyropartitioning of TRUs by chlorination of denitrated HLLW and molten salt/liquid metal extraction.

For this purpose, a new experimental apparatus has been installed at JRC-ITU. It consists of stainless steel box equipped with telemanipulators and double glove box, both operated in a pure Ar atmosphere. The electrorefiner consists of three electrodes and a liquid Cd pool covered by a molten LiCl-KCl eutectic mixture. The steel box will later be installed in a lead shield in order to treat real HLLW and irradiated fuel.

Introduction

The interest in pyrometallurgy is increasing more than ever after the selection a few decades ago of oxide fuel and aqueous reprocessing as the fuel cycle reference. The reason of this interest can be attributed to the drastic change of boundary conditions with regard to the nuclear fuel cycle in the world. The former mission of the fuel cycle was to recover Pu as an important fissile material for the fast breeder reactor, however at present we can hardly expect positive credit from recovered Pu. Today's main emphasis is put on maximal cost reduction for cycle technology. Furthermore, recovery of long-lived nuclides becomes a new requirement, since geological disposal of high-level waste (or once-through fuel) faces large difficulties with regard to public acceptance. Recovery of long-lived nuclides means employment of various reactor systems for transmutation, resulting in new requirements to reprocess diversified fuels, e.g. MOX, metal fuel, nitride fuel, high burn-up fuel, etc. These new requirements may bring us to a different choice for future fuel cycle technology. Pyrometallurgical processing is one of the most attractive alternatives to meet these requirements. The requirement for product purity being much less stringent, the recovery of minor actinides (MA: Np, Am, Cm) will take place simultaneously with plutonium due to the thermodynamic properties of molten salt media. The recovery of MA allows the reduction of TRU wastes, and simultaneously decreases the risk of nuclear proliferation. The molten salt media also provide two important features as a solvent material in nuclear processing. The radiation stability of molten salt will enable us to process spent fuels of high radioactivity (e.g. spent fuel with short cooling time) without any increase of solvent waste. Since molten salt does not work as neutron moderator as water does, a comparatively larger amount of fissile material can be handled in the process equipment. Compact and economical facilities are the expected result of this feature.

The Central Research Institute of Electric Power Industry (CRIEPI) found these promising features in pyroprocess as a result of an information exchange with the US-EPRI in 1985. The feasibility of pyrometallurgy to separate/recover actinides from spent nuclear fuel or high-level liquid waste (HLLW) began in 1986 [1,2,3]. As a joint study with US-DOE, CRIEPI participated in the Integral Fast Reactor (IFR) Programme of Argonne National Laboratory (ANL) from 1989 to 1995 in order to study the development of pyrometallurgical technology [4,5] and to demonstrate the pyroprocess of spent metal fuel [6]. In parallel, the measurement of thermodynamic properties of actinides as well as pyrometallurgical partitioning of TRUs from simulated HLLW had been carried out by CRIEPI in collaboration with Missouri University and Boeing North American [7,8]. Over the course of this study, the feasibility of pyrometallurgical process to recover/separate actinides from spent metal fuel or HLLW was confirmed by the results of experiments with unirradiated TRU materials and theoretical calculation based on measured thermodynamic properties. The demonstration of TRU recovery from spent metal fuel (or even from unirradiated ternary alloy fuel) has, however, been left unrealised because of a sudden cancellation of the IFR programme. The Institute for Transuranium Elements (JRC-ITU) has studied the capacities of the aqueous process for the separation of TRUs from HLLW for many years [10]. The CRIEPI and JRC-ITU collaboration for the study of metal target fuels for transmutation of TRU [9] has led to a new joint study on pyrometallurgical processing. The new joint study will demonstrate the feasibility of pyrometallurgical processes to separate actinide elements from real spent fuel and HLLW, which are the missing keys for a rational evaluation of future fuel cycle technology. Though the joint study has just started, it has already attracted considerable attention from the scientific community. Several organisations have expressed interest in joining the study for closer information exchange. This study has also been included in a project of the European 5th Framework Programme, which CRIEPI has joined as a member without funding. In this paper, the current status as well as the whole test plan of this joint study will be reported.

Test plan

The first phase of the joint study ranges from 1998 to 2004, during which time three different stages are to be carried out, as shown in Figure 1. The first stage is "development and installation of the experimental apparatus". In this stage, a facility consisting of a hot cell and a glove box under argon atmosphere dedicated for pyrometallurgical experiments is developed. The electrorefiner to be used in the subsequent stage is also developed in this period. The second stage is "development of metal fuel reprocessing", during which recovery of actinides from unirradiated metal alloy fuel such as U-Pu-Zr and U-Pu-MA-RE-Zr is to be carried out. The metal alloy fuel is first treated with an electrorefining step described in Figure 2 [2]. After electrorefining, the molten salt electrolyte used in the electrorefiner will be treated in a multi-stage molten salt/liquid metal extraction process for the recovery of residual actinides and keeping sufficient separation from lanthanides. Figure 3 shows the schematic view of the multi-stage extraction step, where counter current contact will be employed [7]. The recovered TRU-Cd metal will be submitted to a distillation step for separating TRU from Cd. The third stage is "demonstration of TRU recovery from HLLW" where pyrometallurgical partitioning is to be demonstrated using actual HLLW. Chlorination, multi-stage extraction and Cd distillation will first be tested with unirradiated materials containing TRU elements. At a later stage, the whole system will be moved inside a 15 cm lead shielding for biological protection in order to carry out the demonstration test described below. The actual HLLW is first converted into oxide form in the chemical cell by condensation and calcination [11]. The oxide products are then converted into chlorides under Cl_2 gas flow at about 973 K. The reaction vessel for the chlorination step where oxides are reacted with Cl_2 gas in molten salt bath is schematically shown in Figure 4. During the chlorination reaction, some volatile chlorides such as Fe_2Cl_6, $ZrCl_4$ and $MoCl_5$ will escape from the molten salt bath. Therefore another molten salt bath at a lower temperature is attached to this chlorination reactor for trapping these volatile chlorides [11]. The obtained chloride salt mixture will be used for the multi-stage extraction process, followed by the Cd distillation step described previously. The treatment of real irradiated metal fuel will be carried out in a subsequent phase.

Development of Ar atmosphere hot cell

The experimental apparatus was newly designed and fabricated for this study (see Photos 1 and 2). It consists of a stainless steel box with telemanipulator and a double glove box, both operated under a very pure Ar gas atmosphere continuously decontaminated by separate purification unit. The requirements for oxygen and moisture concentrations are less than 10 ppm, respectively. Both devices are equipped with heating wells in which high temperature experiments will be carried out. Most of the experiments described in the former section will be carried out in the stainless steel box, whereas preparation and treatment of chemicals/samples as well as basic studies will be carried out in the glove box. The detail of this glove box will be described in another paper at this workshop [12]. The stainless steel box of $1\,600 \times 1\,600 \times 1\,600$ mm has a vertical heating well (ϕ 150 mm in diameter, 600 mm in depth) which consists of an inconel liner and a stainless steel tube. The box is provided with an airlock system for introduction or extraction of items without deteriorating the purity of the Ar atmosphere. A La Calhene container, which can be connected to the air lock system, will be used for the transfer of samples and materials to the glove box or to other hot cells. The well is situated on the bottom of the box, and heated from outside by a cylindrical resistance heater up to 1 273 K. In order to reduce the diffusion of oxygen, double sleeved manipulators with intermediate Ar flushing are employed. Over the course of installing the stainless steel box, many initial problems had to be faced, i.e. leakage at the airtight sealing of the double manipulator sleeve, malfunction of remote operation of the airlock door, etc. After many modifications in-house, both boxes are now in an operational condition.

Development of electrorefiner

The electrorefining step is the key part of pyrometallurgical reprocessing, since fuel dissolution as well as actinides refining take place in this one step. An electrorefiner is newly designed and fabricated in CRIEPI according to the experience accumulated in numerous experiments. It should be noted that the design concept of this electrorefiner is to demonstrate the separation and recovery yield for alloyed fuel reprocessing. The process speed should be demonstrated later by a larger system like the high throughput electrorefiner being developed by ANL [13]. We think the most important thing to demonstrate now is the recovery of actinides from real metal alloy fuel with a sufficient level of decontamination from FP elements, because ANL had to abandon their demonstration test before entering into plutonium operation.

The designed electrorefiner consists of three electrodes and a liquid Cd pool covered by a molten LiCl-KCl eutectic mixture, as schematically shown in Figure 5. It was shipped to JRC-ITU, and installed in the stainless steel box after the modification for manipulator use as shown in Photo 3. The electrorefiner cell of ϕ 100 mm $\times$ 130 mm is hung on a metal flange that has seven holes for cathode, anode, stirrer, reference electrode, sampler, etc. The flange is cooled by an external refrigerator. Fuel alloy rods fabricated at ITU in the framework of a joint study with CRIEPI on the transmutation of TRU will be changed into a steel mesh basket to work as anode. The pool of liquid Cd below molten salt media works as an anode, cathode, or just a receiver for non-reactive elements which fall from the anode basket. The cathode assembly of the electrorefiner uses either a solid iron cathode for U recovery or a liquid metal cathode for TRU recovery. The cathode assembly can be lifted as shown in Photo 4 to be exchanged with another cathode assembly. The solid iron cathode (ϕ 18 mm) with spiral groove will be rotated during electrodeposition to obtain better recovery [2]. On the other hand, the liquid metal itself will be stirred by rotating the ceramic stirrer submerged in liquid metal cathode (surface area = 8 cm^2) in order to avoid the formation of U dendrites that will hamper the deposition of plutonium [2]. About 80 g of metal Cd will be charged in the ceramic cathode crucible. For monitoring the electrode potentials, an Ag/AgCl reference electrode will be employed because of its reliability. The concentration of relevant elements in each phase will be measured by chemical analyses of the samples periodically obtained. The temperature of the electrorefiner will be kept at 773 K during operation.

Experimental strategy of electrorefining

Unirradiated metal fuel will be treated in the first phase experiments to study the unknown behaviour of TRUs during electrorefining as well as the operation sequence to maximise the TRU recovery that will be crucial for the experiments with real irradiated fuel.

The electrorefiner will be loaded with approximately 1 000 g of LiCl-KCl eutectic salt and 500 g of cadmium. The whole system will be heated up to the operation temperature of 773 K to melt both phases. A salt treatment operation to decrease the moisture and oxygen contents in the system will be performed if necessary. Depleted U metal will then be charged to the system followed by addition of CdCl$_2$ to oxidise U metal into UCl$_3$, since some amount of actinide chloride is necessary to facilitate electrotransportation in the molten salt electrolyte. After the equilibrium and the stability of the system are confirmed, the operation with depleted U will be carried out to test the apparatus and the procedure. Polarisation curves will be measured for both anode and cathode against an Ag/AgCl reference electrode. Electrodeposition of uranium metal on the solid iron cathode and the liquid Cd cathode will be tested with a constant current determined from the polarisation curves. Plutonium metal will next be charged in the system to create the salt composition. Unirradiated U-Pu-Zr metal alloy fuels are then charged into the anode basket for anodic dissolution or direct dissolution in the Cd pool.

Plutonium recovery into a liquid Cd cathode will be performed after several fuel treatments to recover only uranium onto the solid iron cathodes. Over the course of this experiment, a suitable dissolution method will be selected from anodic dissolution or direct dissolution. According to the previous experiments [14], the anodic dissolution method may cause complex exchange reactions between alloyed plutonium and UCl_3 in the salt electrolyte, although it has higher dissolution rate than that of the direct dissolution method.

FP simulating elements such as lanthanides will next be added to the system, and electrorefining of U-Pu-MAs-REs-Zr will be performed for simulating the operation of irradiated metal alloy fuel. Fuel dissolution will be performed according to the method determined above. TRU recovery into the liquid Cd cathode will then be carried out after U recovery on the solid cathode. The cathode products will be analysed so as to determine the recovery yield and the decontamination factors.

Conclusions

The present status of the joint study "Small-scale demonstration of pyrometallurgical processing of metal fuel and HLLW" was summarised with a complete experimental plan of the first phase. Despite the relatively small amounts involved, this study will provide the first demonstrative values on the pyrometallurgical processing of HLLW and metal fuels. We believe that the results will give important data for a rational selection of future nuclear options.

REFERENCES

[1]   T. Inoue, T. Sakata, M. Miyashiro, H. Matsumoto, M. Sasahara, N. Yoshiki, "Development of Partitioning and Transmutation Technology for Long-Lived Nuclides", *Nucl. Technol.*, Vol. 93, 206 (1991).

[2]   T. Koyama, M. Iizuka, H. Tanaka, M. Tokiwai, "An Experimental Study of Molten Salt Electrorefining of Uranium Using Solid Iron Cathode and Liquid Cadmium Cathode for Development of Pyrometallurgical Reprocessing", *J. Nucl. Sci. Technol.*, Vol. 34 (4), 384-393 (1997).

[3]   T. Inoue and H. Tanaka, "Recycling of Actinides Produced in LWR and FBR Fuel Cycle by Applying Pyrometallurgical Process", Proc. GLOBAL'97, Yokohama, Japan, 5-10 Oct. 1997.

[4]   Y.I. Chang, "The Integral Fast Reactor", *Nucl. Technol.*, Vol. 88, 129 (1989).

[5]   T. Koyama, T.R. Johnson and D.F. Fischer, "Distribution of Actinides Between Molten Chloride Salt/Liquid Metal Systems", *J. Alloys Comp.*, Vol. 189, 37 (1992).

[6]   M. Lineberry, H.F. McFarlane and R.D. Phipps, "Status of IFR Fuel Cycle Demonstration", Proc. GLOBAL'93, Seattle WA, 12-17 Sept. 1993, p. 1066.

[7]   Y. Sakamura, T. Hijikata, T. Inoue, T.S. Storvick, C.L. Krueger, J.J. Roy, D.L. Grimmet, S.P. Fusselman and R.L. Gay, "Measurement of Standard Potentials of Actinides (U,Np,Pu,Am) in LiCl-KCl Eutectic Salt and Separation of Actinides from Rare Earths by Electrorefinig", *J. Alloy. Comp.*, Vol. 271-273, 592-596 (1998).

[8]   K. Kinoshita, T. Inoue, S.P. Fusselman, D.L. Grimmett, J. Roy, R.L. Gay, C.L. Krueger, C.R. Nabelek and T.S.Storvick, "Separation of Uranium and Transuranic Elements from Rare Earth Elements by Means of Multistage Extraction in LiCl-KCl/Bi System", *J. Nucl. Sci. Technol.*, 36, 189-197 (1999).

[9]   T. Inoue, M. Kurata, L. Koch, J-C. Spirlet, C.T. Walker and C. Sari, "Characterization of Fuel Alloys with Minor Actinides", *Trans. Am. Nucl. Soc.*, Vol. 64, 552 (1991).

[10]   O. Courson, R. Mambeck, G. Pagliosa, K. Roemer, B. Saetmark, J.P. Glatz, P. Baron, C. Madic, "Separation of Minor Actinides from Genuine HLLW Using the DIAMEX Processes", Proc.5th International Information Exchange Meeting, Mol, Belgium, 25-27 Nov. 1998, p. 121.

[11]   K. Kinoshita, M. Kurata, K. Uozumi, T. Inoue, "Estimation of Material Balance in Pyrometallurgical Partitioning Process for TRUs from HLLW", Proc. 5th International Information Exchange Meeting Mol, Belgium, 25-27 Nov. 1998, p. 169.

[12]   C. Pernel, J-P. Glatz, M. Ougier, L. Koch, T. Koyama, "Partitioning of Americium Metal from Rare Earth Fission Products by Electrorefining", Proc. OECD Workshop on Pyrochemical Separation, Avignon, 14-15 March 2000.

[13]   E.C. Gay, W.E. Miller, J.J. Laidler, "Proposed High Throughput Electrorefining Treatment for Spent N-Reactor Fuel", Proc. Embedded Topical Meeting on DOE Spent Nuclear Fuel and Fissile Material Management, Reno, Nevada, 16-20 June 1996, p. 119.

[14]   T. Koyama, M. Iizuka, R. Fujita, "Development of Molten Salt Electrorefining – Direct Electrotransportation from U or Ce to Solid Cathode", CRIEPI-REPORT (in Japanese) T93052, 1994.

Figure 1. Test schedule of joint study (first phase)

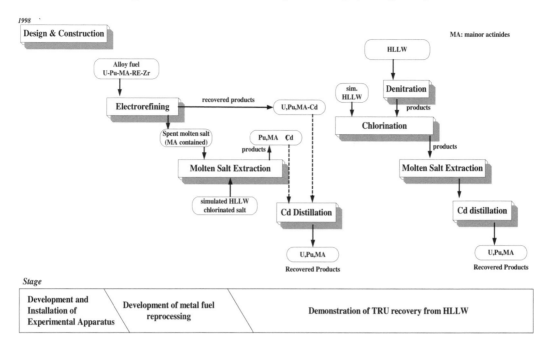

Figure 2. Electrorefining step

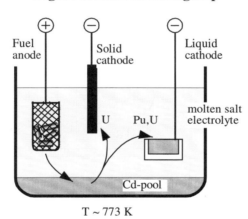

Figure 3. Schematic drawing of counter current multi-stage extraction step

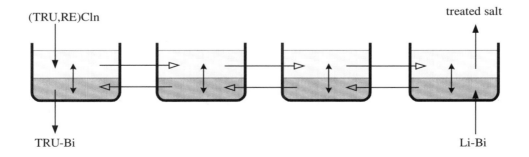

Figure 4. Chlorination step

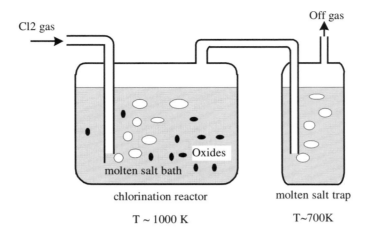

chlorination reactor

T ~ 1000 K

molten salt trap

T~700K

Figure 5. Electrorefiner for metal fuel reprocessing

1) Molten salt electrolyte, 2) Anode basket, 3) Liquid Cd pool, 4) Reference electrode,
5) Ceramic stirrer for liquid metal cathode, 6) Liquid metal cathode, 7) Cylindrical heater,
8) Inner floor of stainless steel box, 9) Heating well, 10) Cooling flange, 11) Cathode lift

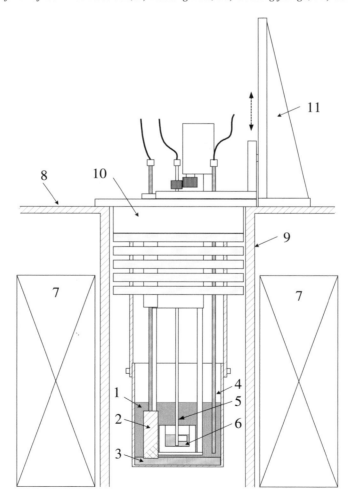

Photo 1. Stainless steel box with Ar purification unit

Photo 2. Glove box with Ar purification unit

Photo 3. The electrorefiner installed in stainless steel box (at lift-up position for photo view)

Photo 4. Electrorefiner (solid cathode assembly is lifted up for viewing)

STUDIES ON THE HEAD-END STEPS FOR PYROCHEMICAL REPROCESSING OF OXIDE FUELS

T. Subramanian, B. Prabhakara Reddy, P. Venkatesh,
R. Kandan, T. Vandarkuzhali, K. Nagarajan and P.R. Vasudeva Rao
Fuel Chemistry Division
Indira Gandhi Centre for Atomic Research
Kalpakkam 603 102, India

Abstract

The molten salt electrorefining process is the most suited pyrochemical process for advanced fuels such as alloys, carbides and nitrides. A laboratory scale facility has been set up for carrying out studies on this process. Studies have been carried out on the recovery of uranium by using molten salt electrorefining. Chlorination of the oxides of uranium to form the oxychlorides is the head-end step of the pyrochemical reprocessing method for oxide fuels. Studies were carried out on the chlorination of uranium oxides at 500°C and on the effect of the presence of carbon on the chlorination.

Introduction

Pyrochemical reprocessing methods offer several advantages over their aqueous counter parts, such as the ability to process short cooled fuels, reduction in doubling time, compact plant, less waste, actinide recycle, etc. [1,2]. In the Fast Breeder Test Reactor at Kalpakkam, uranium-plutonium mixed carbide is being used as the fuel. Molten salt electrorefining, a pyrochemical reprocessing method, besides being the most suited one for the metallic fuels, can be used for other advanced fuels like carbides. Hence studies on molten salt electrorefining have been undertaken in our centre to establish the optimum experimental conditions for various separations. For this purpose, a laboratory scale facility has been set up and has been successfully operated for the last five years. All the steps in the processing flowsheet, such as dissolution of uranium and its alloys in cadmium, separation of U from alloys by electrorefining, consolidation of deposits, distillation of cadmium and injection casting of metallic uranium have been demonstrated. This paper presents a brief description of this lab scale facility and our experience with this facility.

Mixed oxide of uranium and plutonium has been chosen as the fuel for the prototype fast breeder reactor proposed to be set up in our county. Hence studies on pyrochemical reprocessing of oxide fuels have been taken up recently. The pyrochemical reprocessing method for oxide fuels differs from that used for metal alloys especially in the head-end steps. Mixed oxide fuel can be brought to a solution in molten salt through a chlorination process, which has in fact, been established by RIAR, Russia [3]. The uranium and plutonium oxides are dissolved as the oxychlorides in the molten salt medium from which crystalline oxides can be deposited on a carbon cathode. In our laboratory, experiments have been undertaken to study different possible routes for the head-end step. We have studied three possible variants:

- Chlorination of uranium oxides at 500-600°C in molten salt by heating under flowing chlorine gas.

- Direct chlorination of uranium oxides in the solid state under flowing chlorine gas.

- Reaction of uranium oxides with cupric chloride in LiCl-KCl salt at 500°C.

The effect of the presence of carbon on the extent of chlorination has been studied. The preliminary results of these studies are presented and discussed in this paper.

Studies on molten salt electrorefining process for advanced fuels

Laboratory scale facility for studies on molten salt electrorefining

The laboratory scale facility set up [4] and being operated in our laboratory for the studies on the molten salt electrorefining process for metallic fuels is shown in Figure 1. The facility consists of four argon atmosphere glove boxes which house the electrorefining cell, consolidation set up for separation of the uranium from the occluding salt and an injection casting set up for making rods from the metals recovered in the molten salt electrorefining process. The facility is maintained under a negative pressure of 25 mm with respect to atmosphere. The pressure control system is based on neoprene diaphragm operated pressure control valves. For carrying out the preparation of salt mixture, electrorefining, etc., furnaces are attached to the floors of the glove boxes in the facility. The furnaces are of split type whose units can be separated once the experiments are over and the equipment has to be cooled. A 12 kW induction heater is positioned outside the lab scale facility and the coils are taken into the glove box through leak tight fittings. The induction heater is used in the consolidation step for heating the cathode deposit to 1 300°C to separate the uranium metal from the occluding salt as well as for the injection casting of the molten metals. An air atmosphere glove box is also attached to the argon atmosphere glove boxes for housing the cadmium distillation set up.

Electrorefining experiments

A schematic diagram of the electrorefining cell used in our experiments is shown in Figure 2. It consists of an external SS304 vessel in which a SS430 crucible is positioned. The external vessel is fitted with a flange having the provisions for introducing the anode, cathode and a stirrer rod. The SS430 crucible was used for containing the anode, cathode and the electrolyte salt. A stainless steel basket having two decks was used for supporting the anode crucible as well as the cathode rod. In the bottom deck, pieces of uranium metal were kept in a stainless steel crucible and the basket was made the anode. In the upper deck, an alumina crucible was kept surrounding a 12 mm steel rod which was used as the cathode. The alumina crucible served as an insulator as well as a collector for any deposit falling off the cathode. Electrorefining experiments were carried out to determine the extent of recovery of uranium achieved. About 200 g of uranium turnings were taken as the anode and about 3 kg of the LiCl-KCl electrolyte salt in which 6 wt.% UCl_3 has been loaded was used as the electrolyte. When non-perforated crucibles were used for the collection of the deposit, non-uniform deposits of uranium were obtained. In the bottom portion of the cathode rod which was inside the alumina crucible, almost no deposit was present whereas a large amount of deposit was found on the top portion of the rod which was above the collector crucible. Hence a perforated pyrophilite crucible was used for deposit collection when uniform deposit was obtained because free flow of the electrolyte was enabled. The potential of the cell was controlled between 0.8 and 1.0 V Typical current densities were about 50 mA/cm^2. Collection efficiencies were ranging from 60 to 85% in the different runs. Further studies are in progress to improve the collection efficiency.

Studies on chlorination of uranium oxides

A schematic diagram of the chlorination set up used in these experiments for the studies on uranium oxides is shown in Figure 3. The set up is housed in an FRP fume hood. It consists of a one end closed glass tube, closed with suitable fitting on the other end, which has provision for inserting a gas inlet tube. The chlorine gas used for the equilibration was generated *in situ* in another glass vessel by reacting manganese dioxide with hydrochloric acid at 70 to 80°C. The chlorine gas thus generated was passed through a trap containing sulphuric acid to remove any moisture before sending into the reaction tube. The chlorine gas coming out of the reaction vessel was passed through a wash bottle containing sulphuric acid to avoid any back diffusion of moisture. Finally, the gas was passed through a trap containing sodium hydroxide before venting it out.

Chlorination experiments were carried out on UO_2 and U_3O_8 samples with and without graphite powder. In these experiments, about 20 g of uranium oxide was taken in the reaction tube and reacted with chlorine gas, which was passed through the reaction tube for about 12 hours. In order to ascertain the influence of carbon on the reaction rate, in some of the experiments, about 6 g of graphite powder was intimately mixed with the uranium oxide before reacting with chlorine. At the end of the experiment, the reacted sample was removed and a portion was taken for analysis. The analytical sample was dissolved in water and the filtrate was analysed for uranium using the Davis-Gray method [5]. The residue was presumed to be unreacted uranium oxide. Another representative sample was dissolved in phosphoric acid to prevent any possible oxidation of U(IV) to U(VI) and the filtrate was analysed by spectrophotometry for U(IV) and U(VI).

In some of the experiments, UO_2 dispersed in molten LiCl-KCl salt was used for the chlorination studies. In these studies, typically 100 g of salt was dispersed with about 20 g of UO_2. In another experiment, about 100 g of LiCl-KCl molten salt containing 20 g UO_2 was reacted with excess of cupric chloride for about 4 h and the salt was analysed by dissolving in water. The samples were taken periodically to monitor the progress of the reaction.

Results of the studies on chlorination of uranium oxides

The results of the studies on chlorination of uranium oxides at 500°C for 12 hours are given in Table 1. The results given indicate that in the presence of carbon, both UO_2 and U_3O_8 were converted almost quantitatively. When carbon was not present, the rate of conversion and hence the percentage of conversion was lower by almost half in the case of UO_2, whereas almost no conversion occurred in the case of U_3O_8. The results from spectrophotometric analysis are shown in Figure 4. As shown in the figure, only U(VI) was found to be present in the samples obtained by chlorination of UO_2 and U_3O_8 in the presence of graphite. In the case of samples obtained by the chlorination of uranium oxides in the absence of carbon, the spectra show the presence of mainly U(VI) with traces of U(IV). In all these cases, UO_2Cl_2 is mainly formed which is shown by the U(VI) peak in the spectra. In the case of uranium oxides reacted with chlorine in the absence of graphite, a small amount of $UOCl_2$ is also formed which is indicated by the presence of U(IV) peaks in the phosphoric acid solution. Thermodynamic data indicate that formation of UCl_4 is not possible in these two cases.

The results presented in the table also indicate that the rate of conversion is almost double, when the chlorination was done on UO_2 dispersed in molten salt compared to direct chlorination of UO_2 powders. Hence of these two routes, chlorination of the powders dispersed in molten salt may be a preferred head-end step compared to the direct chlorination. Besides, the molten salt route has the added advantage that it can be directly used as the electrolyte for the next step, viz. electrodeposition of the oxides.

It is possible that cupric chloride could aid in the dissolution of uranium oxides, since it could be expected to get reduced to cuprous chloride with simultaneous oxidation of uranium oxide (U(IV)) to uranium oxychloride (U(VI)). The preliminary results obtained on the reaction of UO_2 dispersed in LiCl-KCl with cupric chloride at 500°C were, in fact, comparable to those obtained by the direct chlorination of UO_2 powder. However, further studies are needed to arrive at a firm conclusion.

REFERENCES

[1] L. Burris, R.K. Steunenberg and W.E. Miller, AIChE Symposium Series, 254, 83(1987), 134.

[2] Y.I. Chang, *Nucl. Technol.*, 88 (1989), 129.

[3] A.V. Bychkov, S.K. Vavilov, P.T. Porodnov, O.V. Skiba, G.P. Popkov and A.K. Pravdin, Molten Salt Forum, Vols. 5-6 (1998), 525.

[4] T. Subramanian, B. Prabhakra Reddy, R. Kandan, P. Venkatesh, T. Vandarkuzhali, K. Nagarajan and P.R. Vasudeva Rao, Proc. 5th Inter. Conf. on Recycling, Conditioning and Disposal (RECOD'98), Vol. 3, p. 775 (1998).

[5] W. Davis and W. Gray, *Talanta*,11 (1964), 1203.

Table 1. Results of chlorination experiments on uranium oxides

Temperature: 500°C, Time: 12 h

No.	Sample	Percentage conversion
1	UO_2 (LiCl-KCl)	84
2	UO_2	40
3	U_3O_8	5
4	UO_2 +C	73
5	U_3O_8 + C	88

Figure 1. Laboratory scale facility for electrorefining studies

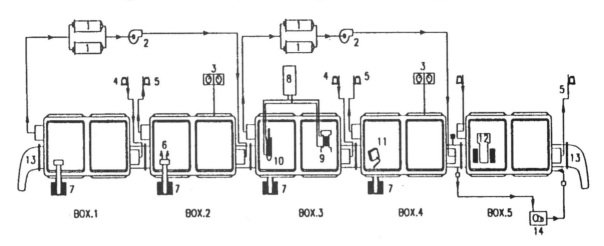

ARGON ATMOSPHERE GLOVE BOXES

BOX.1 ELECTROLYTE SALT PREPARATION
BOX.2 ELECTROREFINING EXPERIMENTS
BOX.3 INDUCTION MELTING
BOX.4 METAL/ALLOY TREATMENT

AIR ATMOSPHERE GLOVE BOX
BOX.5 TRANSIT AIR BOX (CADMIUM DISTILLATION)

1. PURIFICATION TOWER
2. BLOWER
3. PHOTOHELIC GAUGE
4. VACUUM BREAKER (INLET)
5. VACUUM REGULATOR (EXHAUST)
6. ELECTROREFNING CELL
7. GLOVE BOX (FURNACE ATTACHMENT)

8. INDUCTION GENERATOR UNIT
9. CONSOLIDATION SET UP
10. INJECTION CASTING SET UP
11. POLISHING INSTRUMENT
12. CADMIUM DISTILLATION SET UP
13. BAG-IN/BAG-OUT PORT
14. VACUUM PUMP

Figure 2. Schematic diagram of electrorefining cell

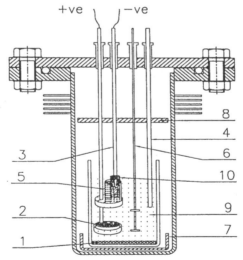

1. LIQUID CADMIUM POOL
2. ANODE
3. CATHODE ROD
4. REFERENCE ELECTRODE
5. PERFORATED PYROPHILITE CRUCIBLE
6. STIRRER
7. CERAMIC INSULATOR
8. HEAT SHIELD
9. MOLTEN SALT
10. CATHODE DEPOSIT

Figure 3. Schematic diagram of salt purification set up

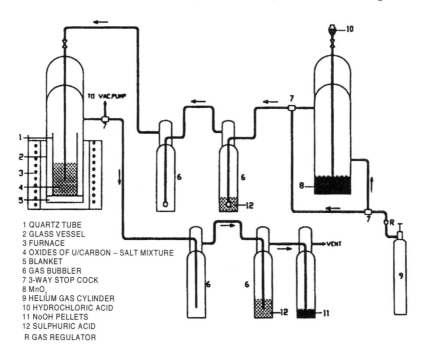

1 QUARTZ TUBE
2 GLASS VESSEL
3 FURNACE
4 OXIDES OF U/CARBON – SALT MIXTURE
5 BLANKET
6 GAS BUBBLER
7 3-WAY STOP COCK
8 MnO_2
9 HELIUM GAS CYLINDER
10 HYDROCHLORIC ACID
11 NoOH PELLETS
12 SULPHURIC ACID
R GAS REGULATOR

Figure 4. Spectra of the chlorinated samples in phosphoric acid solution diluted with water (2:1)

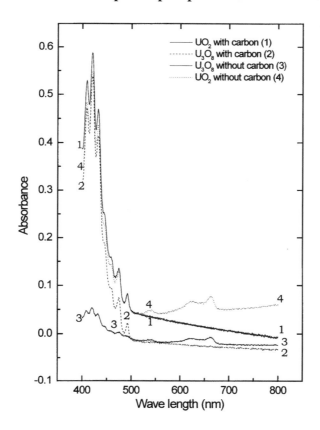

SESSION III

Part C: Process Simulation and Process Design

Chairs: J. Uhlir, H. Boussier

DEVELOPMENT OF STRUCTURAL MATERIAL
AND EQUIPMENT FOR MOLTEN SALT TECHNOLOGY

Pavel Hosnedl
Škoda JS a.s. (Škoda Nuclear Machinery plc)
Orlík 266
316 06 Plzen, Czech Republic

Oldřich Matal
Energovýzkum, Ltd., Božetěchova 17
612 00 Brno, Czech Republic

Abstract

Structural material MoNiCr alloy is an experimental material used by the development of deteriorated nuclear fuel transmutation technology. The basic requirements on the material of reactor vessel with all attachment are corrosion resistivity in the fluoride salts environment and creep strength by reactor working temperature. The experimental material tested showed convenient corrosion and creep properties. Forming technology of this material is to a certain extent difficult, as maintaining specific formability conditions is necessary.

The experimental study of pyrochemical separation processes with regard to future industrial applications requires a variety of technological equipment such as molten salt pumps, pipelines and heat exchangers. This is why development of these components has begun.

In this paper, information is provided on the design of a molten salt loop with forced circulation by a pump.

Introduction

Within the solution of the Czech national project of transmutation our own way of the corrosion resistant alloy development has been chosen. This high nickel content alloy is designated as a structure material for fluoride experimental loops for the operation temperatures of approximately 700°C.

Until now the total number of five half-tonne ingots of high nickel content alloy MONICR ŠKODA have been cast. All recently cast material was re-forged in ŠKODA STEEL to semi-products for the manufacture of sheets, tubes and bars according to the technology developed last year and published [1] and [2].

All semi-products required in the shape of narrow sheet were rolled at the University in Freiberg (Bergakademie Freiberg). Semi-products for all flanges and plugs were forged from the bars and the bars were used for semi-products for the manufacture of tubes of different diameters and for the manufacture of receivers, nozzles and terminal pieces.

Extruding of pipes of OD 36 mm, ID 22 mm and length 6 m was performed in Iron Works Hrádek. During the pressing some problems arose which are usual for the technology of extruding of pipes made of this alloy.

High deformable strength of the material was the reason for the interruption of the forming process at first. During the other experiment the semi-product temperature was elevated and the whole semi-product was pressed.

In addition to the manufacture of semi-products there was a number of material analyses performed the task of which was to determine parameters of the newly developed material.

Within large R&D works recrystallisation properties of the alloy during rolling were checked in the extent of applied rolling temperatures, and material creep characteristics were measured and corrosion experiments in molten fluoride salts in special stands were performed (see pictures in Appendix*).

Recrystallisation of rolled semi-products

Rolled semi-products were subject to large research of the material recrystallisation behaviour during forming at different temperatures and follow-up annealing. Some samples were subject to cold forming. Selected samples, re-forged with consequent rolling to the thickness of 40 mm were subject to experimental treatment at different modes.

After every forming heating, the sample was subject to two-stage deformation the size of which was different for individual modes. The samples heating temperatures before the beginning of forming and actual temperatures during the pass of each sample between the rollers are lower and are different in comparison with the first and second deformation.

These temperatures were measured for each sample by pyrometer. Such treated samples were subject to measurements on an optical microscope and a scanning electron microscope.

Each sample was analysed in the central area .All microstructures on the pictures were consequently evaluated using image analysis system "Image Pro" software.

---

* The Appendix was unavailable at the time of publication.

Due to the used contrast it was possible tell the difference between recrystallised and non-recrystallised structures, the areas were evaluated with the assessment of recrystallisation structure share for each mode.

Measurement of forming forces and alloy deformation resistance

Using force detectors located directly in embedded rollers it was possible to measure radial forces during the rolling process. These forces were consequently recalculated to technological deformation resistance of the alloy.

The Appendix shows the correlation of the deformation resistance as a temperature function. This relation shows logical dependence of the increase of the deformation resistance together with the decreasing temperature.

In global point, it is possible to characterise the deformation resistance in the monitored temperature range as very high resistance, it represents more than threefold strength in comparison with standard quality structure steel.

Measurement of creep and corrosion properties of alloys

Samples for the measurements of creep characteristics were made of material treated by B1, B2, B3 and D1 modes. Two samples were measured for each mode. The measurement was performed at a temperature of 710°C and applied stress of 160 MPa.

The measurement results are indicated in the Appendix. It is obvious there, that creep properties significantly depend on the method of the samples treatment. It is remarkable that the time to rupture increases with the growing deformation strengthening of material.

Corrosion tests in operation medium – fluoride salt – are performed on experimental stands. This will be followed by demanding stress corrosion cracking tests for various conditions of sensitising heat treatment and for weld joints of MONICR material semi-products. The material is in the stage of licensing.

Components of the experimental loop

For experimental purposes test loops with a fluoride molten salt are needed. To design and then operate a test loop many specific components should be available that provide among others:

- Molten salt circulation in the loop (pumps).

- Pipeline opening and/or closure (valves).

- Molten salt flow rate control (control valves).

- Heating of test loop structural materials (heating elements).

- Molten salt melting and heating in tanks (internal heaters).

- Molten salt heating in test loop operation (through-flow heaters).

- Molten salt cooling in test loop operation (heat exchangers).

Basic requirements these components have to meet are:

- Ability to operate in fluoride molten salt environment in temperature range over 400°C in the test loop design life time period.

- Acceptable corrosion resistance of component structural materials in fluoride molten salt under high temperatures.

- To secure staff safety during the test loop operation.

Design principles

Principles of component design issue from our experience gained from development, design manufacture and operation of sodium test loops and lead-bismuth test loop that had been in operation in laboratories in Brno and also from development, design and manufacture of high power level steam generators for sodium cooled fast breeder reactors. For example two of them (each of 30 MW thermal power and sodium inlet temperature of 500°C) are operating at the fast reactor BOR-60 in Russia [3].

Some technical data of components for small loops

Molten salt pump

The molten salt pump is designed as an impeller vertical pump with a flange-mounted electric motor. The pump in the version for small fluoride experimental loops is designed for nominal parameters of molten salt:

- Flow rate of 2.5 kg/s.

- Temperature of 500°C.

- Outlet/ inlet pressure difference of 0.6 MPa.

All parts of the pump which are in contact with the salt are from corrosion resistant materials. The pump design provides also function of a loop buffer tank.

Valves

Both control and closing valves are designed with bellows in I.D. range of 15 to 80. The valve body as well as all valve parts in contact with the fluoride salt are from corresponding corrosion resistance materials.

Molten salt pressure gauges

To measure the pressure of the molten salt pressure gauges with a double barrier between the salt and environment with temperature compensation have been designed. Pressure gauges for small loops are designed for pressure range up to 1 MPa with an analogue output. A temperature sensor is also built in the pressure gauge.

Molten salt storage tanks

Molten salt storage tanks are designed as cylindrical pressure vessels with external and internal located heaters and heating elements and control elements like pressure, level and temperature sensors. Vacuum as well as cover gas systems can bee connected to the storage tank.

Heat exchangers

Heat exchangers for small test loops at the power level of approximately 200 kW are designed in four versions, namely:

- Heat exchanger molten salt – molten salt.

- Heat exchanger molten salt – molten metal.

- Heat exchanger molten salt – water, water steam.

- Heat exchanger molten salt – gas (air).

Heat exchangers in all for versions are designed as tube heat exchangers with solved problems of thermal expansion and transients of operation.

Conclusions

Recrystallisation behavior of material

The share of recrystallised structure increases with higher annealing time which follows hot forming and with that the share of recrystallised structure seems to be higher in transverse cross-section than in longitudinal cross-section.

The difference in hot forming conditions between recrystallisation behaviour and deformation resistance did not show any significant effect on the grain size by comparison of samples with the same annealing time.

Different treatment during the different modes does not cause any significant difference in the recrystallised grain size.

Deformation resistance of material

Deformation resistance of material at high temperatures (800-1 200°C) is high with multiple exceeding of values of any known steels.

Creep behaviour of material and corrosion experiments

The measured creep results are rather surprising. They allow to reach the conclusion that the time to rupture increases together with the growing deformation strengthening of material and vice-versa, during annealing, recovering and ongoing recrystallisation the time decreases. A quite different trend can be seen for most of the steels. The differences obtainable by measurement of creep properties are very significant and important.

The longest measured time of deformation strengthened specimen represents 200% increase in comparison with a specimen with the longest time of recrystallisation annealing. We can assume that further optimising of the material treatment will allow further increase of these characteristics.

Currently, there are long-term corrosion experiments in progress. Short-term experiments demonstrated high corrosion resistance of MONICR material against fluoride salts.

Notes to the manufacture of flat semi-products and tubes

Manufacture of flat semi-products

It is possible to state that the technology of flat semi-products manufacture is mastered. Thanks to systematic co-operation with Technical Universities the rolling technological procedures at different temperatures were tested as it is probable that in the future it will be necessary to use also other technological equipment where there is no possibility of as high heating as in the laboratory rolling mill in Freiberg.

Manufacture of tubes

It was demonstrated that forward extruding is the technology which allows to manufacture tubes of MONICR alloy.

Numerical calculation performed with the Austrian software Deform demonstrates that with the change of geometrical ratio it is possible to achieve significantly lower pressing force which will be easily achievable on an extruding press.

The distribution of the deformation field by the extruding of MONICR alloy tubes is numerically simulated by Deform software.

By this simulation also the distribution of temperature and deformation fields is evident. In the area of transition from the pressing tool to the matrix, i.e. in the area of maximum deformation reforming of material, there is significant increase in temperature.

Manufactured semi-products were subject to successful verification of electron beam welding and welding with filler material. Non-destructive and destructive testing of weld joints documented excellent results meeting strict requirements for nuclear manufacture according to ISO 9001 and ASME Code.

Activities focused on research into and usage of fluoride molten salt for nuclear spent fuel transmutation technologies began in the Czech Republic approximately three years ago. These activities have been also supported by experience gained from development, manufacture and supply of equipment for fluoride technologies (NRI Řež – FREGAT) and for fast sodium cooled reactors (institutes and companies from Brno steam generators for BOR-60 and BN-350).

First steps in R&D of new technologies have been oriented as into theoretical as into experimental analysis at all times.

So we hope our contribution to the international co-operation in the first phase of the fluoride molten salt technology development could be in research and implementation of components for fluoride molten salt experimental and test loops.

REFERENCES

[1]   M. Hron, P. Hosnedl, V. Valenta, "Short Information about Conception and Works for Nuclear Incineration Systems in Czech Republic", IAEA Consult, 27-29 October 1997, Cadarache, France.

[2]   P. Hosnedl, Z. Nový, V. Valenta, "Development of Corrosion Resistance Alloy MONICR SKODA", Proceedings of 3rd Int. Conf. on Accelerator-Driven Transmutation Technologies and Applications, Prague, 7-11 June 1999.

[3]   "Fast Reactor Steam Generators with Sodium on the Tube Side", IAEA-TECDOC-730, IAEA Vienna, January 1994.

[4]   F. Dubšek, O. Matal, "Steam Generators and Intermediate Heat Exchangers for a Molten Salt Transmutor Demo Unit", Proceedings of 3rd Int. Conf. on Accelerator-Driven Transmutation Technologies and Applications, Prague, 7-11 June 1999.

KINETICS OF PYROCHEMICAL PROCESSING OF NUCLEAR MATERIALS USING MOLTEN SALT AND LIQUID METAL

Hirotake Moriyama, Hajimu Yamana
Research Reactor Institute, Kyoto University
Kumatori-cho, Sennan-gun, Osaka, 590-0494, Japan

Kimikazu Moritani
Department of Nuclear Engineering, Kyoto University
Yoshida, Sakyo-ku, Kyoto, 606-8501, Japan

Abstract

In support of the development of a metal transfer process, a kinetic aspect of reductive extraction using molten salt and liquid metal is discussed by considering recent experimental results. In the case of a mechanically agitated two-phase contactor system, a critical problem is recognised for the kinetics in which the overall mass transfer coefficients is much lower than would be expected. The problem is explained by considering that the rate is limited by the solubility of the solute metal elements in the metal phase. Some suggestions are given for the further development of similar extraction systems.

Introduction

Pyrochemical processing of radioactive materials using molten salt and liquid metal as solvents is expected to be of much use for advanced nuclear chemical processing. In fact, this kind of processing is indispensable for the reprocessing of nuclear fuels of metallic fuelled fast breeder reactors and of molten salt breeder reactors. Also, because of its high radiation resistance, compactness and less waste generation, it calls much attention for its possible use in place of the present aqueous reprocessing of oxide fuels.

Extensive studies have been performed on the thermodynamics of pyrochemical processing but little is still known concerning the kinetics. In the case of molten salt breeder reactor (MSBR), for example, a mechanically agitated two-phase extractor system was suggested in which the intermediate product of [233]Pa and the neutron poisons of rare earth fission products were removed from the fuel salt into liquid bismuth [1,2]. However, a critical problem was recognised for the kinetics in which the overall mass transfer coefficients were much lower than would be expected, and hardly any meaningful correlations were observed between the overall mass transfer coefficient and the agitation speed [3].

In the present study, a kinetic aspect of pyrochemical processing is discussed by considering recent experimental results. The rate of reductive extraction of actinide and lanthanide elements has been determined in the two-phase system of molten fluoride and liquid bismuth [4], and it has been found that the solubility of those elements in the metal phase is very important for limiting the rate of extraction. To confirm this theory, the diffusion coefficients of the dissolved species have also been measured in each phase [5,6]. By combining these observations, the mechanism of reductive extraction and the performance of extractors are discussed for further development of the extraction system.

Metal transfer process for MSBR

In the conceptual design of MSBR developed at ORNL [1,2], molten fluoride mixtures are used as the fuel carrier and coolant. The fuel salt must be reprocessed continuously in order to meet a high breeding ratio. The main functions of the reprocessing are to isolate [233]Pa from the region of high neutron flux and to remove the rare earth fission products of high neutron absorption cross-sections. The processing method involves the reductive extraction of these components from the fuel salt into liquid bismuth solutions in a two-phase contacting systems.

Mechanically agitated non-dispersing two-phase extractors have been suggested for reductive extraction with the aim to eliminate possible corrosion problems by entrained bismuth, and engineering tests using such extractors with salt and bismuth flow rates (~1% of those required for processing the fuel salt from a 1 000 MW(e) MSBR) have been conducted (1) to study the process itself, (2) to measure the removal rate of representative rare earth fission products from MSBR fuel salt, and (3) to evaluate the mechanically agitated contactor for use in a plant processing fuel salt from a 1 000 MW(e) MSBR [3].

It was successfully shown in the tests that representative rare earth fission products were extracted from the MSBR fuel salt and selectively transferred into liquid bismuth in the stripper vessel with the separation factors of about 10^4 to 10^6 with respect to thorium. However, it was also shown that overall mass transfer coefficients measured at the salt-bismuth interface in the process were much lower than would be required ($>10^{-4}$ m/s) for full-scale metal transfer process equipment of reasonable size. Furthermore, any meaningful correlations were hardly observed between the overall mass transfer coefficients and the agitation speed. Although no detail was known for the mechanism, it was suggested to increase agitation to the point of some degree of dispersion of the salt and bismuth in the stripper.

Mass transfer coefficient

It is important to study a detailed mechanism of reductive extraction for increasing the rate of extraction. Following some studies on the thermodynamics of the extraction process [7], the rate of reductive extraction of actinide and lanthanide elements was measured in the two-phase system of molten fluoride salt and liquid bismuth under a natural convection [4]. In the measurement, it was observed that the rate of accumulation of solute elements in the bismuth phase was not equal to the rate of depletion in the salt phase, and that the material balance in both phases was broken for several hours after the addition of reductant to the system. The formation of bismuthides of solute elements at the salt side interface was inferred from the observation and the following extraction mechanism was presented:

$$Li_{m-n}MF_m(salt) + n\ Li(Bi) + y\ Bi(Bi) \rightarrow MBi_y(interface) + m\ LiF(salt) \tag{1}$$

$$MBi_y(interface) + (x - y)\ Bi(Bi) \rightarrow MBi_x(Bi) \tag{2}$$

Considering Reactions (1) and (2), the kinetic equations are given by:

$$\frac{dC_{M(salt)}}{dt} = \frac{K_{M(salt)}A}{V_{(salt)}}\left(C_{M(salt)} - C'_{M(salt)}\right) \tag{3}$$

$$\frac{dC_{M(Bi)}}{dt} = \frac{K_{M(Bi)}A}{V_{(Bi)}}\left(C'_{M(Bi)} - C_{M(Bi)}\right) \tag{4}$$

where $K_{M(salt)}$ and $K_{M(Bi)}$ are the mass transfer coefficients of the element M in the salt and bismuth phases, respectively, $C_{M(salt)}$ and $C_{M(Bi)}$ the bulk concentrations, $C'_{M(salt)}$ and $C'_{M(Bi)}$ the interface concentrations, A the contact area, and $V_{(salt)}$ and $V_{(Bi)}$ the volumes.

The mass transfer coefficient was evaluated by applying Eqs. (3) and (4) to the experimental data and was interpreted by traditional theories. According to Higbie [8], the mass transfer coefficient K_M is correlated with the diffusion coefficient D_M as:

$$K_M = cD_M^{1/2} \tag{5}$$

where c denotes the proportional constant at given flow conditions. For the diffusion coefficient, the following Stokes-Einstein equation can be applied [9]:

$$D_M = kT/(\alpha\mu R) \tag{6}$$

where k is the Boltzmann's constant, T the temperature, α the constant ($4\pi \leq \alpha \leq 6\pi$), μ the solvent viscosity, and R the radius of the diffusing species. Thus we obtain:

$$K_M = c[kT/\alpha\mu R]^{1/2} \tag{7}$$

The experimentally obtained mass transfer coefficients are well explained by Eq. (7) and, neglecting small differences in R for species in each phase, are roughly expressed as [4]:

$$K_{M(salt)}(m/s) = 5 \times 10^{-8} \left(T/\mu_{(salt)}\right)^{1/2} \tag{8}$$

$$K_{M(Bi)}(m/s) = 6 \times 10^{-9} \left(T/\mu_{(Bi)}\right)^{1/2} \tag{9}$$

where μ is the viscosity in mPa•s. Under the condition of natural convection at 873 K, the $K_{M(salt)}$ and $K_{M(Bi)}$ values are 4×10^{-6} and 2×10^{-6} m/s, respectively.

Diffusion coefficient

Comparing Eqs. (8) and (9), it is found that the mass transfer coefficient values in the metal phase are much smaller than those in the salt phase, and that the radius of the diffusing species in the metal phase will be much larger than that in the salt phase. For confirmation the measurement of the diffusion coefficient of dissolved species has been performed in both phases of Li_2BeF_4 [5] and Bi [6].

For molten salts, Eq. (6) is known to give a good correlation between the diffusion coefficient and viscosity. The viscosity of molten Li_2BeF_4 as a function of temperature [10] can be represented by:

$$\mu_{(salt)}(mPa \bullet s) = 7.4 \times 10^{-4} \exp(4300/T) \tag{10}$$

and by combining Eqs. (6) and (10) we obtain:

$$D_{M[salt]} = c'T \exp(-4300/T) \tag{11}$$

where c' is a constant. The temperature dependence of the measured diffusion coefficients was found to be explained by Eq. (11) and the c' values were determined for some typical ions as shown in Table 1.

Although a reasonable agreement was observed for the temperature dependence, some disagreements between the observation and theory were found for the effective radius of the diffusing species. By taking the α value of 4.6π for molten nitrates [11], for example, the effective radius of the Zr^{4+} species is evaluated from the experimental data to be 0.4×10^{-10} m, which is small compared with the ionic radius of 0.81×10^{-10} m for Zr^{4+}. Considering the microscopic structures of the solvent, the observed discrepancy is attributed to a difference of the local viscosity in comparison with the measured bulk viscosity, and some preferential path may be present for the diffusing species in such a microscopically non-homogeneous structure. For practical applications with molten Li_2BeF_4, this fact is important because rapid kinetics is expected in spite of the high viscosity of the solvent.

The diffusion coefficients of the dissolved species were also measured in liquid Bi [6]. In this case, empirical equations for the diffusion coefficients were given as shown in Table 2. Although no detail of the mechanism has been discussed yet, a similar analysis as above may be applied to the measured data with the viscosity data [12] and the α value around 4.6π, making the R values comparable with those of metallic atoms. Thus the results suggest the diffusion of completely dissolved species in this case.

The results of the diffusion coefficient measurement are compared with the observations in the mass transfer coefficient measurement in which the radius of the diffusing species in the metal phase will be much larger than that in the salt phase. Then it is concluded that the bismuthides of solute elements, which are formed at the salt side interface possibly in the form of some particles of a relatively large radius, will diffuse and dissolve into the metal phase, and that this step will be the rate limiting in the reductive extraction. The problem, in which the overall mass transfer coefficients are considerably low and are poorly correlated with the agitation speed [3], can be explained by considering the formation and slow dissolution of the bismuthides.

Considerations for extractors

For improving the situation, we may recall that the salt-side mass transfer coefficient is rather independent of the formation of the bismuthides. In fact, the salt-side mass transfer coefficients which have been determined by measuring metal concentrations in the salt phase are better correlated with the agitation speed and are much higher than the overall ones [13]. Accordingly the salt-side mass transfer coefficient may be increased to a required level by increasing the agitation speed, although the separation or settlement of the bismuthides will be needed additionally. A direction for the further development of equipments is thus suggested.

In order to ensure the technical feasibility of the metal transfer process, a simple analysis is given here for its performance. For a metal transfer process in which salt and metal of equal flow rates are contacted, the relationship between the flow rate, F, of a fuel salt and the removal time, T_r, of a solute element is given by:

$$F = \frac{V(D_v \eta \varepsilon + 1)}{T_r D_v \eta \varepsilon} \tag{12}$$

where V is the total volume of a fuel salt in primary system, D_v the volumetric distribution coefficient, η the extraction efficiency that accounts for non-equilibrium distribution during contacting and ε the efficiency of back extraction.

Figure 1 shows the dependence of the flow rate on the extraction efficiency for a reference scheme with a removal time of 10 days, in which the following parameter values are taken: $V = 60$ m^3 [1,2], $\varepsilon = 0.99$, and $D_v = 0.1$, 1.0 and 10. As seen in the figure, the flow rate is sensitive to the extraction efficiency in the region of the lower η values, but not elsewhere. The minimum extraction efficiency around 0.5 will be suggested, and the flow rate will be kept below 0.3 m^3/h when $D_v > 10$. Thus, it is capable to meet the removal time of 10 days by using ordinary type extractors of reasonable size.

The extraction efficiency is inherently related with the extractor type and the operating conditions. In order to know the requirements of this factor, an analysis is also performed. By using Eqs. (3) and (4) with the salt-side mass transfer coefficient and by regarding the η as the yield of mass transfer during contacting, we can obtain the following relationship between the η and the contact area, A:

$$\eta = \frac{1 - \exp\left[-(1 + D_v)K_{M(salt)}A / (D_v F)\right]}{1 + D_v \exp\left[-(1 + D_v)K_{M(salt)}A / (D_v F)\right]} \tag{13}$$

Eq. (13) can be combined with Eq. (12) to obtain the dependence of the flow rate, F, on the capacity factor, $K_{M(salt)}A$, as shown in Figure 2. It is found that the flow rate will be kept below 0.3 m^3/h when $D_v > 10$ and $K_{M(salt)}A > 2 \times 10^{-4}$ m^3/s. Such a condition may be satisfied with the use of the mixer/settler type extractors by increasing the agitation speed.

Conclusions

For the development of a metal transfer process, a kinetic aspect of reductive extraction using molten salt and liquid metal was discussed by considering recent experimental results. In the case of a mechanically agitated two-phase contactor system for MSBR, a critical problem was recognised for the slow kinetics in which the rate is limited by the solubility of the solute metal elements in the metal phase. Considering that the salt-side mass transfer coefficient is independent of the slow dissolution, however, it is feasible to increase the salt-side mass transfer coefficient to the required level by increasing the agitation speed.

Similar extraction systems have also been suggested for the reprocessing of metallic fuels, the group partitioning of radioactive wastes, and the tritium recovery from fusion reactor blanket materials. The presently obtained results may be instructive for the development of these systems.

REFERENCES

[1]   A.M. Weinberg, *et al.*, *Nucl. Appl. Technol.* 8 (1970), 105.

[2]   "Conceptual Design Study of a Single-Fluid Molten Salt Breeder Reactor", R.C. Robertson, ed., ORNL-4541 (1971).

[3]   H.C. Savage and J.R. Hightower, Jr., "Engineering Tests of the Metal Transfer Process for Extraction of Rare Earth Fission Products from a Molten Salt Breeder Reactor Fuel Salt", ORNL-5176 (1977).

[4]   H. Moriyama, M. Miyazaki, Y. Asaoka, K. Moritani and J. Oishi, *J. Nucl. Mater.*, 182 (1991), 113.

[5]   H. Moriyama, K. Moritani and Y. Ito, *J. Chem. Eng. Data*, 39 (1994), 147.

[6]   H. Moriyama, Y. Asaoka, K. Moritani and Y. Ito, "Extraction Rate of Metal Elements in Molten Salt and Liquid Metal Binary System, (III) Diffusion Coefficient", presented at the annual mtg. of Atomic Energy Society of Japan, Osaka, 28-30 March 1991.

[7]   J. Oishi, H. Moriyama, K. Moritani, S. Maeda, M. Miyazaki and Y. Asaoka, *J. Nucl. Mater.*, 154 (1988), 163.

[8]   R.B. Bird, W.E. Stewart and E.N. Lightfoot, "Transport Phenomena", Wiley, New York, 1960.

[9]   H. Lamb, "Hydrodynamics", Cambridge University Press, London & New York, 1932.

[10]   S. Cantor, W.T. Ward and C.T. Moynihan, *J. Chem. Phys.*, 50 (1969), 2874.

[11]   S. Forcheri and V. Wagner, *Z. Naturforsch.*, 22a (1967) 1171.

[12]   "CRC Handbook of Chemistry and Physics, 64th Edition", R.C. Weast, M.J. Astle and W.H. Beyer, eds., CRC, Boca Raton, 1983-1984.

[13]   C.H. Brown, Jr., J.R. Hightower, Jr. and J.A. Klein, "Measurement of Mass Transfer Coefficients in a Mechanically Agitated, Nondispersing Contactor Operating with a Molten Mixture of LiF-BeF2-ThF4 and Molten Bismuth", ORNL-5143 (1976).

Table 1. Diffusion coefficients of some typical ions in molten Li_2BeF_4 [5]

Ion	T (K)	$D_{M(salt)} = c'T \exp(-4\,300/T)$; c' $(m^2/s/K)$
Np^{3+}	813-1 023	$(5.7 \pm 0.2) \times 10^{-10}$
Pa^{4+}	813-1 023	$(4.7 \pm 0.4) \times 10^{-10}$
Eu^{2+}	873-1 023	$(6.9 \pm 0.2) \times 10^{-10}$
Ce^{3+}	813-983	$(5.2 \pm 0.2) \times 10^{-10}$
La^{3+}	813-1 023	$(5.8 \pm 0.2) \times 10^{-10}$
Zr^{4+}	873-983	$(3.2 \pm 0.1) \times 10^{-10}$

Table 2. Diffusion coefficients of dissolved species in liquid Bi [6]

Species	T (K)	$D_{M(Bi)} = D_0 \exp(-E/RT)$	
		$\log D_0$ (m^2/s)	E (kJ/mol)
Np	530-710	-7.43 ± 0.21	13.2 ± 3.6
Eu	530-710	-7.24 ± 0.72	14.7 ± 13.0
Ce	620-710	-7.23 ± 0.15	16.7 ± 2.5
La	530-710	-7.22 ± 0.13	16.8 ± 2.3
Zr	530-710	-7.46 ± 0.33	15.5 ± 5.6

Figure 1. Fuel salt flow rate required for the removal time of 10 d as a function of extraction efficiency (see Eq. (12) in text)

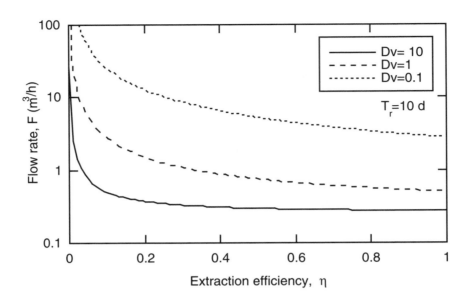

Figure 2. Fuel salt flow rate required for the removal time of 10 d as a function of capacity factor

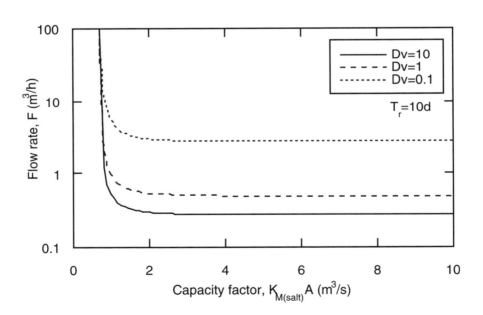

A RATIONAL APPROACH TO PYROCHEMICAL
PROCESSES VIA THERMODYNAMIC ANALYSIS

Gérard Picard, Sylvie Sanchez
Laboratoire d'Électrochimie et de Chimie Analytique, Unité Mixte du CNRS n°7575
École Nationale Supérieure de Chimie de Paris
11 rue Pierre et Marie Curie, 75231 Paris Cedex 05, France

David Lambertin, Jérôme Lacquement
Laboratoire d'Étude des Procédés Pyrochimiques, DCC/DRRV/SPHA
Bâtiment 166 Atalante – CEA Marcoule
BP 171, 30207 Bagnols sur Cèze, France

Abstract

Prediction of selective separation pyrochemical processes can be made from thermodynamical data. In order to illustrate the methodology to be used (based on the smallest number of parameters to be acquired), we shall consider the important case of actinide/lanthanide separation. More particularly, the chemical stabilities of plutonium and cerium trichlorides and oxide compounds in two molten chloride mixtures of different intrinsic acidities (LiCl+KCl eutectic melt and NaCl+KCl equimolar mixture) – whose common working temperatures are respectively 723 K and 1 000 K – will be examined.

Introduction

The feasibility of a new process – or the improvement of a current one – should firstly be fundamentally examined from a thermodynamic point of view. This means that the nature and composition of the solvent melt as well as its working temperature are chosen with respect to the temperature dependence of the chemical and electrochemical properties both of the elements to be treated and of the chemical reagents to be used.

The second investigation step concerns the kinetics of the reactions involved which can modify the choice of melt and temperature in order to have the best compromise for fast and selective chemical reactions. This study also gives us information on the type of reaction (gas-solid, gas-liquid or liquid-liquid reactions).

The ultimate fundamental investigation step is a correct understanding of the reaction mechanisms at the atomic level (nanochemistry), and preliminary results and trends of our recent approach in that field based on molecular modelling techniques were given [1-15].

This paper is centred on the first step that is a rational thermodynamic assessment of the potentialities of molten salt solution chemistry to perform selective separation of lanthanides and actinides and to produce pure actinide oxides.

In order to describe our approach we have chosen to illustrate our purpose by considering the oxide chemistry in molten chlorides, which is an eminently important aspect of French nuclear activity. Obviously, this approach is easily transposable on the one hand to the chemistry of other compounds (e.g. nitrides) and on the other hand to other types of melts. In addition, complexation reactions can provide another determining action mean (e.g. competitive ion exchange reactions between fluoride and chloride anions in chloro-fluoride mixtures). This last aspect will be set aside in this paper. The target elements we have considered are cerium and plutonium. Finally, the description of our methodology will be based on equilibrium data coming from G.S. Picard, *et al* [16] and M.N. Levelut [17] for cerium, respectively in molten LiCl+KCl eutectic and in equimolar NaCl+KCl mixture, and from J. Roy, *et al* [18] and V. Silin and Skiba [19] for plutonium data in those salts. No attempt to make a critical analysis of the available literature data has been made; the reader is invited to consult other papers dealing with this subject at this workshop.

Our research strategy

Molten salts are reaction media which can be used advantageously compared with the more classical media such aqueous and organic solvents. They present some undoubted advantages compared to the classical reaction media, not only by their physical properties, but also and especially by their particular chemistry which authorised for example the processing of very reactive metals.

Yet, they are often used taking advantage exclusively of their physico-chemical properties (ionic liquids with high electrical conductivities, specific solvation powers, large domain of melting points and stabilities, and so on) but few in widely using the richness of solution chemistry (acidity, redox, complexation reactions) which has made the success of aqueous solutions.

When we wish to conceive a new process, in a new kind of solvent, we face a difficult question that needs to be answered: How to choose the best salt to obtain the best selectivity and efficiency for a given process?

This is a fundamental point to work in melts, because we have not only one reaction medium (as water), but a great number of possible solvents. To a great extent – independently of the abundance of this reaction medium – the success of aqueous chemistry is due to the fact that we have at our disposal a great number of data that we can manipulate in order to conceive some selective chemical reactions. In contrast, this is not generally the case in melts. Moreover due to their multiplicity it is not envisaged to experimentally establish them as was done in water systems by generations of scientific researchers. So the great number of melts which theoretically allows to choose the best salt for a given process, is a serious drawback when compared to water. Therefore we need predictive techniques for lowering the number of experiments and to compare the various melts and possibilities depending on their various chemical properties over a wide working temperature range and to establish the required minimum number of basic data upon which a rational approach can be made.

This question gives rise to another set of questions: (i) How to characterise the solvent melt properties regarding elements and chemical reagents under consideration?, (ii) Is it possible to separate the investigations of the chemical behaviours of the products to be treated from those of the reactants?, (iii) Is it possible to compare them in various molten salts and several temperatures?

Positive answers to these questions can be made. We need only to have some selected basic data (experimentally determined or predicted ones) and a universal reference system [20].

About oxide chemistry in molten chlorides

Oxoacidity concept

Molten salt chemistry is now widely investigated using the same concepts as in aqueous media. Solution chemistry is based on particle exchange reactions which allow us to define acidity (if the particle p is H^+, we have the well known Bronsted acidity for aqueous solutions), complexation (p = ligand) and oxydoreduction (p = e^-) concepts via acceptor-donor couples as follows:

$$Acceptor + p = Donor$$

Defining melt acidity from the exchange of the proton really has a sense only for room temperature molten salts such as organic salts (e.g. ethylammonium chloride or ethylpyridinium bromide [21]) or low melting inorganic melts (such as hydrogenosulfates [22]). In contrast, at high temperatures most of the Bronsted acids undergo a decomposition reaction (excepted for gaseous hydrogen chloride) and the Bronsted acidity scale would be very short, indeed punctual. On the contrary for high temperatures, oxides play an analogous role as the acids (or basis) for low temperatures. Then, it is quite obvious to define an acidity (called oxoacidity) from the oxide anion exchange reaction:

$$Acceptor + O^{2-} = Donor$$

The oxide anion being a basic particle, the acceptor of O^{2-} is called an oxoacid and the donor an oxobase.

Oxoacidity level in chloride melts

Local or conditional oxoacidity – definition and measurement

From the preceding definition, the oxoacidity can be quantified by the cologarithm of the oxide anion activity, or in most cases by the cologarithm of its concentration $pO^{2-} = -\log O^{2-}$. This is a conditional acidity scale, that is to say, which depends on melt composition and temperature. Such a scale enables us to make comparisons of the oxoacidity properties of various compounds only in a same melt and at the same temperature. In order not to be limited by this constraint, we have proposed to use an oxoacidity function [20].

Oxoacidity function or universal acidity scale

The function which we have suggested is based on the oxoacidic properties of well-determined gaseous molecular compounds and so can be also related to experiments whatever their temperatures. This function previously described [20] is based on the chemical action of hydrogen chloride onto the oxide anion to give rise to the evolution of gaseous water:

$$O^{2-} + 2HCl(g) = H_2O(g) + 2Cl^- \tag{1}$$

The well known entity pO^{2-} which characterises the oxoacidity in melt is then connected to the cologarithm of the equilibrium acidity constant of Reaction (1):

$$pO^{2-}(T,m) = pK(T,m) + \log [P(HCl)^2/P(H_2O)] \tag{2}$$

where $pK(T,m)$, and m stand respectively for the cologarithm of the equilibrium constant of Reaction (1), temperature and melt composition. We note that contrary to pO^{2-} and $pK(T,m)$ values which depend on melt composition and temperature, this is not the case for the term involving the partial pressures of $HCl(g)$ and $H_2O(g)$ whose value can in fact be fixed independently of the working temperature and melt composition.

Thus we can rewrite Eq. (2) as:

$$pO^{2-}(T,m) = pK(T,m) + \Omega \tag{3}$$

which defines the oxoacidity function Ω as:

$$\Omega = \log [P(HCl)^2/P(H_2O)] \tag{4}$$

allowing us to compare the acidity levels of various melts whatever the temperature may be. The knowledge of the equilibrium constant $pK(T,m)$ of Eq. (2) allows us to calibrate the conditional pO^{2-} scale with respect to the universal oxoacidity scale or Ω scale. Values of $pK(T,m)$ are available in the literature for different molten salts [20].

Redox properties

For characterising the potential independently of temperature, the logarithm of the chlorine partial pressure has to be retained, as can easily be seen from the following equation:

$$Eeq = E^\circ [Cl_2(1\ atm)/Cl^-)] + (2.303\ RT/2F) \log [P(Cl_2)/atm]$$

where Eeq and $E^\circ [Cl_2(1\ atm)/Cl^-)]$ stand respectively for the equilibrium potential and the standard potential of the chlorine (1 atm)/choride electrochemical system. Taking this redox system as a potential reference, we simply have:

$$\log [P(Cl_2)/atm] = Eeq /(2.303\ RT/2F)$$

Thus the literature data generally available under the form of conditional potential-acidity diagrams can be easily transformed into log $[P(Cl_2)/atm]$-Ω equilibrium diagrams.

Temperature dependence

It is now possible – in order to analyse and compare reaction chemistry in molten chlorides at various temperatures – to construct universal three-dimensional diagrams [20] using the reciprocal value of temperature as the third parameter. This was chosen because the pK values of chemical equilibria generally vary as a function of $1/T(K)$. We then define a gaseous partial pressures/temperature reference system which might be called Generalised Pourbaix type diagrams (GPTD). We will see that the projections onto the base of this reference system are quite a useful tool for our initial purpose.

Equilibrium reactivity diagrams for the more common gaseous reactants

In connection with the most promising applications of molten salts that are selective chlorination and selective precipitation of metal oxides, we have established the GPTD diagrams for the more common gaseous reactants for performing the oxide ion exchange reactions given in Table 1. Most of them are very efficient, in particular those concerning an oxidant and a reducing agent such as chlorine and carbon respectively. By combining these reactions with that relative to the action of HCl onto the oxide anion, we can obtain the oxide exchange reactions whose the acidity constant can be calculated from tabulated data.

The reactivity domains of chlorine + oxygen, hydrogen chloride + water + hydrogen and chlorine + carbon oxide gaseous mixtures are reported in Figure 1 as well as the comparison of their chlorinating or precipitating powers (Figure 2), putting in evidence a loss of selectivity when temperature increases (the reactivity domains move towards each other).

Chemical properties of cerium and plutonium in molten chlorides

Our thermodynamic approach was previously developed for the treatment of raw materials, such as in extractive metallurgy of aluminium (selective chlorination of ores such as bauxites [23-25] and ilmenite [26] or the treatment of industrial concentrates (recovery of metals from the treatment of lead and zinc concentrates [27]). Recently we applied our research strategy to the chemical treatment of nuclear wastes [28].

Accurate thermochemical data on metal oxides are often available in the literature. These data coupled with experimental values of standard potentials for the relative metal/metal chloride electrochemical systems provide us data on the stabilities of metal oxides in melts. Unfortunately, for most cases, data concerning pure metal oxychlorides are not available. Moreover practically no solubility data of oxychlorides were determined. Therefore, we have to determine the stability of these species using either specific probes such as the yttria-stabilised zirconia membrane electrode to experimentally measure the oxoacidity (acidobasic potentiometric titrations [29]), or to perform molecular modelling calculations (previously developed). In the following we will illustrate our approach by considering the processing of oxide compounds in molten chlorides.

Cerium and plutonium chemical properties in LiCl+KCl eutectic melt

Electrochemical window

The 2-D diagrams represented in Figure 3 give the available potential-oxoacidity domain in the molten LiCl+KCl eutectic as a function of temperature. Because of dissociation processes, the electrochemical window decreases by about 30% ($\log P(Cl_2)$/atm varies from -50 to -35 for the reduction limit of the solvent melt).

Cerium properties

Cerium is much more acidic at high temperature (1 000 K) than at low temperature. In fact the cerium oxychloride precipitates from a solution of cerium trichloride 1 $mol.kg^{-1}$ at an oxoacidity value Ω of -1.82 at 1 000 K and we have to reach a Ω of value of -3.36 for this precipitation at 723 K.

Considering the oxychloride, we can easily see on these diagrams that the stability domain of CeOCl is very large at 723 K. It ranges from Ω = -3.36 to Ω = -10.3. Hence it is very difficult to reach the formation of cerium sexquioxide: $\Delta\Omega$ < -10, that is pO^{2-} (conditional acidity value) practically equal to zero.

In contrast for a higher temperature (1 000 K), the stability domain of cerium oxychloride decreases to the benefit of that of cerium sexquioxide.

Plutonium properties

Plutonium trichloride is also more acidic at high temperature than at low temperature. In both cases, however, the plutonium oxychloride maintains a low stability.

Cerium and plutonium chemical properties in the NaCl+KCl equimolar mixture

The main differences observed in the NaCl+KCl mixture compared with LiCl+KCl melt is the fact that cerium oxychloride is very soluble and that plutonium oxychloride is totally unstable (Figure 4).

Effect of melt composition and temperature on the possibilities for a chemical separation of cerium and plutonium

LiCl+KCl melt

We observe that a separation can be envisaged in LiCl+KCl (Figure 5) because plutonium oxychloride and oxide precipitate before cerium oxychloride. Thus a process can be considered either by selective dissolution of cerium from an oxide mixture or by selective precipitation of plutonium oxide from a given halides mixture. These separations are better at high temperature. At 1 000 K, the predicted efficiency is at best 99% ($\Delta\Omega$ = 2).

NaCl+KCl mixture

Figure 6 demonstrates the influence of the solvation properties of melt on the separation of cerium from plutonium. In fact, we can see that in NaCl+KCl at 1 000 K, due to the fact that cerium oxychloride is highly solvated (soluble species) we can dispose of a domain of acidity of about 6. So, separation processes on the same basis as before (precipitation of plutonium oxide or solubilisation of cerium oxide) is predicted with a very high efficiency ($\Delta\Omega = 6$). If we envisage to electrochemically extract cerium from a mixed solution, however, it is preferable to fix the oxoacidity in the range where cerium trichloride is stable; we dispose then of a $\Delta\Omega$ value of 2.9 and the separation could be reached at about 1%.

Choice of selected chemical reagent as a function of their chlorinating and oxidising powers

To illustrate this aspect we will only consider the actions of Cl_2/O_2 and $HCl/H_2O/H_2$ gaseous mixtures but this is easily transposable to the chemical actions of other mixtures. The reactivity domains of these gaseous reagents are given in Figure 7 (partial pressures of gases ranging from 10^{-3} to 1 atm). First of all, we can see that HCl(g) has a higher chlorinating power that chlorine and can chlorinate all the oxides and oxychlorides considered here. On the contrary, chlorine can only chlorinate cerium oxides in NaCl+KCl at 1 000 K. Obviously these diagrams enable us also to predict the precipitations of pure oxides and it is possible to selectively obtain them with a given oxidation state for cerium or plutonium.

Conclusion

Thermodynamic data allow us to set up generalised Pourbaix type diagrams (GPTD). These 3-D reaction diagrams (potential-acidity-temperature) give reactivity zones relative to the precipitation and solubilisation (eventually followed by a subsequent vaporisation) of metal oxychlorides and oxides. A comparison, depending on melt compositions and working temperatures, of these diagrams enables us to derive the outlines of a pyrochemical separation process.

In addition to acidity and redox effects, we can use complexation reactions to obtain a better selectivity. This is certainly an important aspect for actinides and lanthanides having an intermediate oxidation state.

Such an analytical approach is a very general way (applicable not only to oxides, but also to nitrides, or alloys, molten salt chemistry) for conceiving the skeletal structure of a new pyrochemical process.

REFERENCES

[1]   L. Mouron, S. Grandjean, J-J. Legendre, G. Picard, "A New Method for the Structural Modelling of Disordered Compounds. Application to Molten NaCl", *Computers and Chemistry*, (1994), 18 (1), 5-11.

[2]   G. Picard, F. Bouyer, "Molecular Modelling: An Analytical Tool with a Predictive Character for Investigating Reactivity in Molten Salt Media", *J. Amer. Inst. of Phys.*, 1995, 330, 295-304.

[3]   P. Hébant, G. Picard, "Equilibrium Reactions Between Molecular and Ionic Species in Pure Molten LiCl and LiCl+KCl (M = Na, K, Rb) Melts Investigated by Computational Chemistry", *J. Molecular Structure (THEOCHEM)*, 1995, 358, 39-50.

[4]   L. Mouron, G. Roullet, J-J. Legendre, G. Picard, "Geometrical Analysis of the Voids in Structural Models of Molten Salts", *Computers and Chemistry*, (1996), 20 (2), 227-233.

[5]   F. Bouyer, G. Picard, J-J. Legendre, "Computational Chemistry: A Way to Reach Spectroscopic and Thermodynamic Data for Exotic Compounds", *J. Chem. Information and Computer Sciences*, 1996, 36 (4), 684-693.

[6]   G. Picard, F. Bouyer, M. Leroy, Y. Bertaud, S. Bouvet, "Structure of Oxyfluoroaluminates in Molten Cryolite-Alumina Mixtures Investigated by DFT-Based Calculations", *J. Molecular Structure (THEOCHEM)*, 1996, 368 (1-3), 67-80.

[7]   P. Hébant, G. Picard, "Quantum Mechanics Applied to the Study of (Li,K)X Binary Melts Ionic Dissociation (X = F, Cl, Br or I)", *J. Electrochem. Soc.*, 1997, 144 (3), 980-84.

[8]   L. Mouron, G. Roullet, J-J. Legendre, G. Picard, "Computational Evidence of Complexes in Structural Models of Melts Deriving from Neutron Diffraction Data", *Computers and Chemistry*, (1997), 21 (6), 431-435.

[9]   A. Hemery, P. Hébant, G. Picard, M. Sibony, B. Champin, "Analysis of Gas Phase Chemical Reactions Using Thermochemical Data from Ab Initio Calculations. The Case of Refractory Metals", *J. Molecular Structure (THEOCHEM)*, 1998, 425, 1-12.

[10] L. Joubert, G. Picard, J-J. Legendre, "Structural and Thermochemical Ab Initio Studies of Lanthanide Trihalide Molecules with Pseudopotentials", *Inorg. Chem.*, 37 (1998), 1984-1991.

[11] P. Hébant, G.S. Picard, "Computational Investigations of the Liquid Lithium/(LiCl-KCl Eutectic Melt) Interface", *J. Molecular Structure* (THEOCHEM), 426 (1998), 225-232.

[12] L. Joubert, G. Picard, J-J. Legendre, "Advantages and Drawbacks of the Quantum Chemistry Methodology in Predicting the Thermochemical Data of Lanthanide Trihalide Molecules", *J. Alloys and Compounds.*, 275-277 (1998), 934-939.

[13]   L. Joubert, G. Picard, B. Silvi, F. Fuster "Electron Localization Function View of Bonding in Selected Aluminum Fluoride Molecules", *J. Molecular Structure (THEOCHEM)*, 463 (1999), 75-80.

[14]   L. Joubert, G. Picard, B. Silvi, F. Fuster "Topological Analysis of the Electron Localization Function: A Help for Understanding the Complex Structure of Cryolitic Melts", *J. Electrochem. Soc.*, 146 (6), (1999), 2180-2183.

[15]   L. Joubert, B. Silvi, G. Picard, "Topological Approach in the Structural and Bonding Characterization of Lanthanide Trihalide Molecules", *Theor. Chem. Acc.*, 104 (2000), 109-115.

[16]   G.S. Picard, Y.E. Mottot, B.L. Tremillon, "Acidic and Redox Properties of Some Lanthanide Ions in Molten LiCl+KCl Eutectic, The Electrochemical Society Softbound Proceedings Series, Vol. 86-1 (1986), 189-204.

[17]   M.N. Levelut, thèse Université Pierre et Marie Curie, Paris, 1977.

[18]   J.J. Roy, L.F. Grantham, D.L. Grimmett, S.P. Fusselman, C.L. Krueger, T.S. Storvick, T. Inoue, Y. Sakamura, N. Takahashi, "Thermodynamic Properties of U, Np, Pu, and Am in Molten LiCl+KCl Eutectic and Liquid Cadmium", *J. Electrochem. Soc.*, 143 (8), (1996), 2487-2492.

[19]   V. Silin, O. Skiba, Report NIIAR-P-118 (1971).

[20]   G.S. Picard, "Three-Dimensional Reactivity Diagrams for Reaction Chemistry in Molten Chlorides", *Molten Salt Forum*, Vol. 1-2 (1993-94), 25-40.

[21]   G.S. Picard, J. Vedel, "Chemical Properties in Molten Ethylpyridinium Bromide and Ethylammonium Chloride. II – Interpretation of Halide Activity Coefficients by the Hard and Soft Acids and Bases Concept of Pearson. Application to the Study of Oxide Exchange Reactions", *J. Chim. Phys. Chim. Biol.*, 1975, 72-(6), 767-777.

[22]   J.P. Vilaverde, G.S. Picard, J. Vedel, "Measurement of the pH and Autoprotolysis in Molten Potassium Bisulfate at 220°", *J. Electroanal. Chem. Interfacial Electrochem.*, 1974, 54 (2), 279-88.

[23]   G.S. Picard, F.M. Séon, B.L. Trémillon, "Prediction of the Selective Chlorination of Oxides by Gaseous Mixtures in Molten Lithium Chloride-Potassium Chloride Eutectic", in "Molten Salts", M. Blander, D.S. Newman, M-L. Saboungi, G. Mamantov and K. Johnson, eds., *The Electrochemical Society Softbound Proceedings Series*, Vol. 84-2 (1984), 694-706.

[24]   B.L. Tremillon, G.S. Picard, "Chemical Solubilization of Metal Oxides and Sulfides in Chloride Melts by Means of Chlorination Agents", in "Molten Salt Chemistry – An Introduction and Selected Applications", G. Mamantov and R. Marassi, eds., NATO ASI Series, Ser. C. Mathematical and Physical Sciences, Vol. 202 , D. Reidel Publishing Company (1987), pp. 305, 327 (Proc. of the Advanced Study Institute on Molten Salt Chemistry, Camerino, Italy, 3-15 August 1986 ,under the auspices of NATO and the University of Camerino).

[25]   F. Séon, G.S. Picard, B. Tremillon, Y. Bertaud, *Fr Patent*, 2 514 028 (1981).

[26]   D. Ferry, G.S. Picard, B. Tremillon, *Trans. Inst. Min. Metall.*, Sect. C, 97 (1988), C21-C30.

[27] M. Garcia, Y. Castrillejo, P. Pasquier, G.S. Picard, *Molten Salt Forum*, 1-2 (1993-94), 47-56.

[28] D. Lambertin, J. Lacquement, S. Sanchez, G.S. Picard, "Electrochemical Properties of Plutonium in Molten CaCl₂-NaCl at 550°C", *this meeting*.

[29] G.S. Picard, F. Séon, B. Tremillon, "Oxoacidity Reactions in Molten Lithium Chloride+Potassium Chloride Eutectic (at 470°C): Potentiometric Study of the Equilibria of Exchange of Oxide Ion Between Aluminum(III) Systems and Carbonate and Water Systems", *J. Electroanal. Chem. Interfacial Electrochem.*, 102 (1) (1979), 65-75.

Figure 1. GPTD diagrams for chlorinating gaseous mixtures

Temperature dependence of the reactivity domains of Cl₂ + O₂ (left), HCl + H₂O + H (middle) and CO + Cl₂ gaseous mixtures (right) in molten chlorides (partial pressures ranging from 10³ to 1 atm)

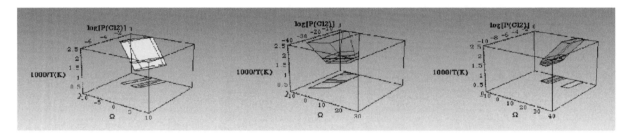

Figure 2. Comparison of the reactivity domains of the main chlorinating gaseous mixtures

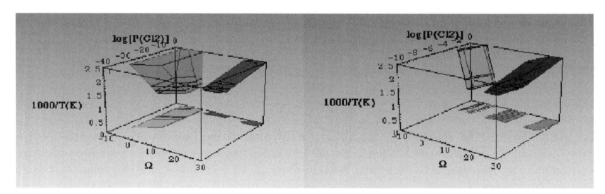

Figure 3. GPTD 2-D diagrams summarising the chemical properties of cerium and plutonium in the LiCl+KCl eutectic melt at 723 K and 1 000 K (molality scale)

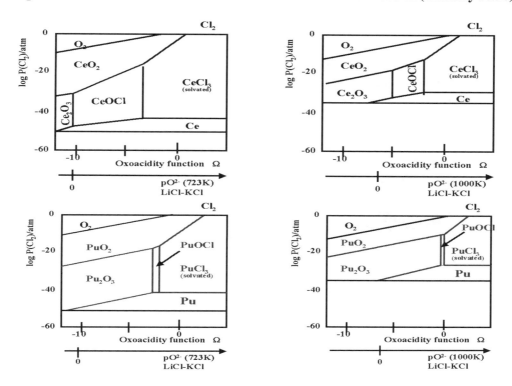

Figure 4. GPTD 2-D diagrams summarising the chemical properties of cerium and plutonium in the equimolar NaCl+KCl melt at 1 000 K (molality scale)

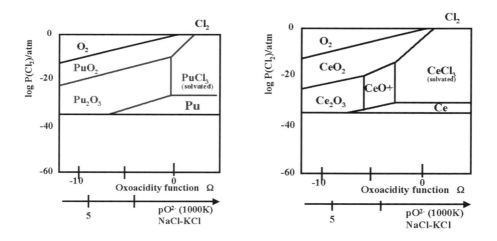

Figure 5. Possibilities for a selective separation of cerium and plutonium in the LiCl+KCl eutectic melt as a function of temperature (723 and 1 000 K)

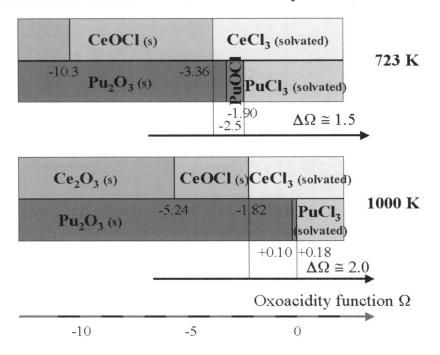

Figure 6. Possibilities for a selective separation of cerium and plutonium at 1 000 K as a function of melt composition (LiCl+KCl or NaCl+KCl)

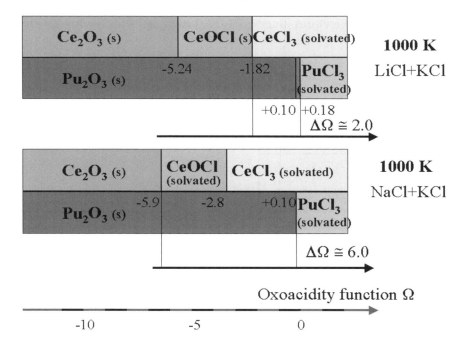

Figure 7. Chemical actions of chlorine-oxygen and hydrogen chloride-water-hydrogen gaseous mixtures on cerium and plutonium species in molten LiCl+KCl eutectic and NaCl+KCl equimolar mixture respectively at 723 and 1 000 K

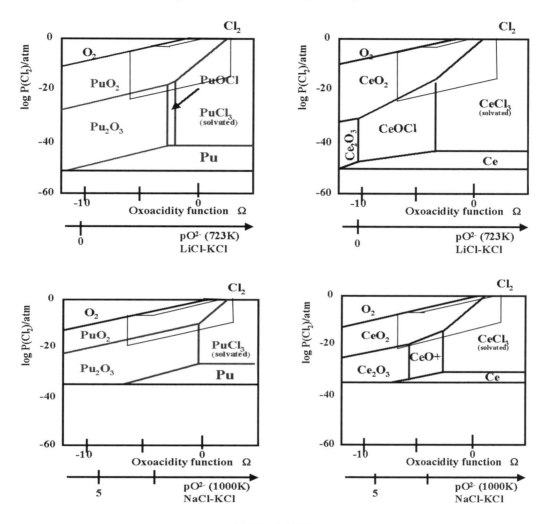

Table 1 [20]

Chemical equilibria	Cologarithms of the constants $pK = A + B \times 1\ 000/T(K)$	
	A	**B**
1a) $Cl_2(g) + H_2O(g) = 0.5\ O_2(g) + 2HCl(g)$	-3.5072	+3.0413
2a) $Cl_2(g) + CO(g) + H_2O(g) = CO_2(g) + 2HCl(g)$	+1.0678	-11.760
3a) $COCl_2(g) + H_2O(g) = CO_2(g) + 2HCl(g)$	-6.0264	-6.0649
4a) $Cl_2(g) + 0.5\ C(s) + H_2O(g) = 0.5\ CO_2(g) + 2HCl(g)$	-3.5546	-7.2502
5a) $Cl_2(g) + C(s) + H_2O(g) = CO(g) + 2HCl(g)$	-8.1769	-2.7408
6a) $CO(g) + H_2O(g) = CO_2(g) + H_2(g)$	+1.8476	-2.0147
7a) $0.5\ CCl_4(g) + H_2O(g) = 0.5\ CO_2(g) + 2HCl(g)$	-7.0566	-4.7011
8a) $SOCl_2(g) + H_2O(g) = SO_2(g) + 2HCl(g)$	-6.3943	-1.4735
9a) $0.25\ S_2Cl_2(g) + 0.75\ Cl_2(g) + H_2O(g) = 0.5\ SO_2(g) + 2HCl\ (g)$	-3.1284	-4.5457

SESSION IV

Open Discussion and Recommendations

Chair: C. Madic

OPEN DISCUSSION AND RECOMMENDATIONS

Professor Madic, the workshop chairman, briefly introduced a proposal for setting up a task force under the auspices of the NEA Nuclear Science Committee for the preparation of a state-of-the-art report on pyrochemical separations (see Annex 3). The floor was then open for discussion.

The participants expressed strong interest in the proposal. The support was mainly due to the following reasons.

- Much work on pyrochemical separations technology has been done in the USA, Europe and Asia. This knowledge should be preserved in order to avoid duplication of efforts and repetitive mistakes, considering the imminent retirement of many of the people who have performed the work.

- Each country has its own future fuel cycle option scenarios. During the next five years, many of these countries will review their research programmes, and a report, furnishing timely and coherent coverage of the situation in each country and providing a common scientific base on pyrochemical separations, would be very useful for promoting efficient international collaboration.

The following comments were made on the scope and objectives of the proposed state-of-the-art report. The report would:

- Examine possible nuclear fuel cycle scenarios within the next 50 years. Future system concepts would be described, including the various reactor types being considered in each country.

- Include applications, advantages and disadvantages of pyrochemical separation from the viewpoint of partitioning and transmutation, as well as for the optimisation of the fuel cycle. Main target areas and specific benefits of pyrochemical separation would be highlighted.

- Cover fission product separation (Cs, Sr, Tc, I, etc.), an issue that has engendered interest from authorities concerned with public acceptance of nuclear power and nuclear waste disposal.

- Cover industrial work, as well as basic research. The extensive experience outside the field of nuclear energy should be taken into account.

- Provide a time plan for future R&D, subjects to be covered and means to solve them.

Participants, especially from Belgium, the Czech Republic, Japan, France, the Russian Federation, the UK and the USA, suggested possible names for participation in the task force. The European Commission would support European participation in the task force. Possible co-operation with the IAEA will be investigated. It was agreed that the task force would consist of both researchers and process engineers. Dr. Laidler was proposed as the chairman of the task force.

Professor Madic concluded from the above discussions that a large majority of the audience was very supportive of establishing a NEA task force for preparing a state-of-the-art report on pyrochemical separations. The proposal to set up a task force would be presented for approval at the next meeting of the NEA Nuclear Science Committee in the beginning of June 2000. Professor Madic mentioned that the detailed scope and objectives would be further discussed at the first meeting of the task force, scheduled for the autumn of 2000.

The necessity for regular workshops on the subject of pyrochemical separations was also discussed. Professor Madic proposed that the next workshop on pyrochemical separations be held after completion of the state-of-the-art report. This proposal was approved by the participants.

POSTER SESSION

TECHNETIUM METAL AND PYROMETALLURGICALLY FORMED SEDIMENTS: STUDY AND SPECIATION BY Tc NMR AND EXAFS/XANES

K. Guerman, T. Reich, C. Sergeant,
R. Ortega, V. Tarasov, M. Simonoff
UMR 5084/CNRS
Université Bordeaux
1, Le Haut Vigneau, Gradignan
ESRF, ROBL, Grenoble/Rossendorf
Institute of Physical Chemistry
Institute of General and Inorganic Chemistry, R.A.S.

Abstract

EXAFS and NMR are being developed with respect to study and speciation of technetium-99 during pyrometallurgical reprocessing of spent nuclear fuel. Based on the set of standards including technetium metal, different technetium halogenides, oxohalogenides, oxides and pyrometallurgically formed sediments, Tc chemical forms responsible for Tc accumulation in fused salts, deposits or gas-off depending on the applied reprocessing conditions are studied by means of NMR and EXAFS. NMR spectrum of Tc metal powder obtained by FT of a free induction decay accumulated after excitation of the spin system by a sequence of high frequency pulses (0.8 μs) with a dead time of 5.4 μs and a repetition time of 1 μs, provided with Knight shift $K(ppm) = 7\,305 - 1.52 \times T$; $\nu_Q(^{99}Tc) = 230$ kHz at 293 K, $C_Q(^{99}Tc) = 5.52$ MHz. EXAF spectra provided with excellent evidence for Tc(IV) halogenides, but need further development for lower oxidation states of technetium.

Introduction

EXAFS and NMR are promising methods to enable us with speciation of radioactive nuclides. It is important to use them for Tc speciation in pyrometallurgy, which is now considered as the most probable alternative approach for reprocessing of spent nuclear fuel. Pyrometallurgy of uranium and plutonium in chloride and fluoride fused salts is now intensively studied, while little attention is paid to long-lived fission product technetium which is one of the hazardous nuclear wastes. During pyrometallurgical spent fuel reprocessing, technetium can remain in the fused salt or enter (depending on the applied temperatures, reagents and potentials) either the sedimented phase contaminating Pu enriched phase, or electrodeposited U phase (see Figure 1). Even by means of X-ray diffraction, Tc could not be detected either in the sediment or in the deposit, the Tc being finely dispersed and the X-ray amorphous. It is therefore very important to develop other methods of analyses such as EXAFS and NMR, which are applicable to Tc speciation in such samples.

Our preliminary results show that under certain conditions Tc can also form several oxides, oxychlorides and chlorides of varying but rather high volatility, turning on an important polluting risk allied to gas-off.

The data on EXAF and NMR spectra of technetium being fragmentary and not presenting the whole of the species possible under pyrometallurgical conditions, it is greatly important to carry out synthetic work. In this way it will be possible to supply a large number of technetium compounds in closed containers which will meet the security and quality demands, radioactivity level and special requirements with regard to sample size, thickness and sample homogeneity.

Some EXAFS measurements of technetium solutions applicable in some content to pyrometallurgcal treatment of spent nuclear fuel were presented earlier by [1,2]. This study was performed with EXAF data on technetium(IV) hexachloride in HCl solution. Recent studies, however, have given some indication of the presence of an unknown Tc chloride complex [3]

Tc metal and its binary alloys with vanadium has been studied earlier by ^{99}Tc NMR at ambient temperature [4,5,6]. In these early works, the measurements were made in relatively low magnetic fields ($B_0 \leq 1.4$ T) in the differential passing mode with field sweep, which made the measurements of the anisotropic Knight shift (K_{an}) difficult. Moreover, the reported values of isotropic Knight shift (K) for metal Tc varied from K = 6 100 ppm in [4, 5] to K = 7 150 ppm and K_{an} = 1 160 ppm in [6].

Experimental

In this work EXAFS measurements are made at Radiochemical Hutch of ROBL/ESRF: beam size of 3×20 mm^2, integrated flux at sample 6×1 011/s 200 mA at 20 kEV, spectral range 5-35 keV and 2/3 filling mode. Beam line control carried out with VME, SUN workstation, SPEC hot cells are used for sample positioning.

[99]Tc NMR spectra of Tc metal powder at 120-400 K, technetium foil and some preliminary results on the K(^{99}Tc) shifts in the sediment samples formed in fused NaCl-KCl at 550°C were recorded on a Bruker MSL-300 radiospectrometer in 7.04 T magnetic field at 67.55 MHz resonance frequency.

Spectra NMR were obtained by FT of a free induction decay accumulated after excitation of the spin system by a sequence of high frequency pulses (0.8 μs) with a dead time of 5.4 μs and a repetition time of 1 μs. The Knight shifts were measured relative to the external standard 0.1 M KTcO$_4$ aqueous solution.

Chemical preparations

All reagents were chemically pure grade. Technetium was purchased as NH_4TcO_4 from Amersham Co. and as $KTcO_4$ from V/O ISOTOP, RF and converted to $HTcO_4$ by cation exchange. $N(CH_3)_4TcO_4$ was prepared as described in Ref. [7], and Tc metal was prepared by reduction of tetramethylammonium pertechnetate $N(CH_3)_4TcO_4$ (of 99.9% purity) in a flow of non-explosive gas mixture 6% H_2 in Ar in a quartz tube at 1 150 K. Samples were cooled, crashed and the 80-150 μm fraction was sieved of for NMR measurements. The same method was used for preparation of the dispersed metal on metal oxide supports MgO, Al_2O_3, modified for evaporation of $N(CH_3)_4TcO_4$ at the later surface. To improve the resolution and accumulation NMR mode minimising the electric contacts, some technetium metal samples were dispersed in a dodecane/apiezon (20:1) mixture and then air-dried.

Tc halogenides and oxides were prepared as described in [8,9]. The reference technetium compounds prepared for the EXAFS study include: chlorides (M_2TcCl_6, M = Li, Na, K, R_4N); fluorides (M_2TcF_6, M = H, Na, K); bromides (M = Na, K, R_4N); oxides TcO_2, Tc_4O_5 and pertechnetates $MTcO_4$ (M = Na, K, Cs). Samples of simulated pyrometallurgically reprocessed nuclear fuel (to be prepared): Technetium hexachlorides in fused salts (chlorides and fluorides); Tc sublimated fraction; Tc precipitated/deposit fraction (all with ^{99}Tc as the only radionuclide).

Results and discussion

The ^{99}Tc NMR spectrum of polycrystalline metallic Tc is comprised of eight symmetrically arranged satellites and the central component whose shape is governed by the Knight shift anisotropy (Figure 3).

The distance in KHz between the first inner satellites is the lowest frequency of a quadrupole transition ν_Q. The line width of the central component at half-maximum was ~ 22 kHz at 293 K thus having masked the effects of shift-anisotropy on the line shape of the central transition ($-1/2 \leftrightarrow +1/2$). At 400 K the width of the line is reduced and is shown as an axially-symmetric line shape with $K_\parallel - K_\perp = -809 \pm 30$ ppm.

Figure 3 presents the temperature dependence of the isotropic Knight shift for technetium metal powder, and Figure 4 presents the temperature independence of the ν_Q for the same sample. This study was provided with $K(ppm) = 7\ 305 - 1.52 \times T$; $R^2 = 0.99$. Experimental $\nu_Q([99]Tc)$ is 230 kHz at 293 K which gives $C_Q([99]Tc) = 5.52$ MHz. The latter is related to electric field gradient (EFG) (q^{exp}) by the equation: $q^{exp}\ [cm^{-3}] = 2.873*10^{22}\ C_Q[MHz]/Q[barn] = 52.86*10^{22}\ cm^{-3}$, where Q = 0.3 barn.

In general, the total EFG at a nuclear site in a noncubic metal results from the addition of two EFG parts, one being due to the lattice of the ionic core (enhanced by the Sternheimer anti-shielding effect) and the other being due to the non-s-character of the conducting electrons:

$$q^{exp} = (1 - \gamma)q^{lat} + q^{el}$$

In the case of Tc, the contribution arising from the ionic core lattice may be interpolated from de Wette's [10] results for $\alpha = c/a = 1.604$. We find that $(1 - \gamma)q^{lat} = 27.57*10^{22}\ cm^{-3}$ (at $\gamma = -5$ [11]). This value is 52% of the experimental value obtained in the present work. This proportion is typical of the most metals in which the $(1 - \gamma)q^{lat}$ contribution is usually lower (in magnitude) than the observed one.

At the ^{99}Tc NMR spectrum of Tc metal foil of 20 μm (Figure 6), the central component is very close to its position in the Tc metal powder sample, but the eight satellites are missed or not clear due to highly defected crystal cell caused by mutual consecutive mechanic treatments. NMR and EXAFS studies of dispersed Tc (supported on oxide matrix of MgO and Al$_2$O$_3$) and of Tc-Ru and Tc-Rh alloys are in progress. The NMR spectrum of the sample, sedimented from K$_2$TcCl$_6$ in a NaCl-KCl melt at 550°C under reducing conditions formed by Ar-1%H$_2$ bubbling, gave evidence for some metal state (9 884 ppm) but also some to 30% of Tc(VII) (31 ppm).

Three main aspects under consideration with respect to EXAFS/XANES studies are as follows:

1) The ligand influence on the technetium EXAF spectra within the same Tc oxidation state (F, Cl, Br, I and OH, H$_2$O being the ligands of interest with respect to pyrometallurgy as a reprocessing method).

2) The oxidation state including the lower and mixed oxidation states as well as Tc cluster formation on the technetium EXAF spectra.

3) Cations on the technetium EXAF spectrum within the same Tc oxidation state (Li, Na, K, Rb, Cs, Me$_4$N, Et$_4$N, Pr$_4$N, Bu$_4$N could be the cations of interest, while TcO$_4^-$ and TcHal$_6^{-2}$ would be the anions under study (some preliminary results are given in Figures 7 and 8).

Tc K-edge k^3-weighted EXAFS and corresponding Fourier transform of the sample of standard Tc(IV) halogenide (Me$_4$N)$_2$TcBr$_6$ is present at Figure 7. This resulted in the following EXAFS structural parameters for Tc-Br:

Tc-Br	N = 5.8(2)	R = 2.51(2) Å	σ^2 = 0.0040 Å^2	ΔE_0 = -16.9(5) eV

Six halogenide atoms and the distance found correspond well to the same parameters known for similar compounds based on X-ray monocrystal analyses.

Tc K-edge k^3-weighted EXAFS and corresponding Fourier transform of the standard Tc(2,5+) halogenide sample K$_3$Tc$_2$Cl$_8$ is present in Figure 8. The first co-ordination sphere of Tc in this compound includes one Tc atom and four Cl atoms. EXAFS structural parameters for K$_3$Tc$_2$Cl$_8$ (in such preliminary approximation) resulted in the following values:

Tc-Tc	N = 1.6(3)	R = 2.20(2) Å	σ^2 = 0.0069 Å^2	ΔE_0 = -1.1(9) eV
Tc-Cl	N = 2.2(4)	R = 2.46(2) Å	σ^2 = 0.0107 Å^2	

The overestimation of Tc-Tc intervention is evidently due to the neglect of the four chlorine atoms bounded to the second technetium atom. This correction of the model will be as soon as possible.

The results show excellent agreement in case of octahedral Tc(IV) bromide, while for Tc cluster compound an important difference between the experimental and the fit lead to the supposition that some more chloride atoms – but not only the first co-ordination sphere chlorides – are influencing the EXAF spectrum.

REFERENCES

[1] K. Ben Said, M. Fattahi, Cl. Musikas, R. Revel, "Spéciation de Tc(IV) dans les solutions chlorées déterminée par spectroscopie d'absorption X (EXAFS)", Comptes rendus de Journées PRACTICE 1998, Villeneuve-les-Avignon, 25-26 Feb. 1999.

[2] R. Revel, C. Den Auwer, "Spectroscopie d'absorption X", Comptes rendus de Journées PRACTICE 1998, Villeneuve-les-Avignon, 25-26 Feb. 1999.

[3] L. Vichot, M.M. Fattahi, C. Musikas, B. Grambov, "Etude des formes réduites du technétium en milieu sulfato-chlore", Comptes rendus de Journées PRACTICE 1999, Villeneuve-les-Avignon, 17-18 Feb. 2000.

[4] W.H. Jones, F. Melford, *Phys. Rev.*, Vol. 125, N. 4, p. 1253 (1962).

[5] D.O. Van Ostenburg, D. Lamm, H.D. Trapp, D.E. MacLead, *Phys. Rev.*, Vol. 128, N. 4, p. 1550 (1962).

[6] D.O. Van Ostenburg, H. Trapp, D. Lamm, *Phys. Rev.*, Vol. 126, N. 3 p. 938 (1962).

[7] K.E. Guerman, A.F. Kuzina, M.S. Grigoriev, B.G. Gulev, V.I. Spitzyn, "Synthesis and Crystal Structure of Me_4NTcO_4", Proc. Acad. Sci. Russ., Vol. 287, N. 3, pp. 650-653 (1986).

[8] K.E. Guerman, M.S. Grigor'ev, A.F. Kuzina, V.I. Spitsyn, "The Structure and Various Physicochemical Properties of Tetraalkylammonium Pertechnetates", *Russ. Journ. Inorg. Chem.*, Vol. 32, N. 5, pp. 667-670 (Engl. transl.); in Russian, pp.1089-1095 (1987).

[9] V.N. Gerasimov, S.V. Kryutchkov, K.E. Guerman, V.M. Kulakov, A.F. Kuzina, "X-Ray Photoelectron Study of Structure of Technetium Compounds", in "Technetium and Rhenium in Chemistry and Nuclear Medicine", Vol. 3. M. Nicolini, G. Bandoly, U. Mazzi, eds., New York, Raven Press, 1990. pp. 231-252.

[10] F.W. De Wette, *Phys. Rev.*, Vol. 123, N. 1, p. 103 (1961).

[11] V.P. Tarasov, S.A. Petrushin, V.I. Privalov, *et al.*, *Koord. Khim.*, Vol .12, N. 9, p. 713 (1986) (English transl.).

Scheme 1. Possible Tc species in pyrochemical reprocessing of spent nuclear fuel

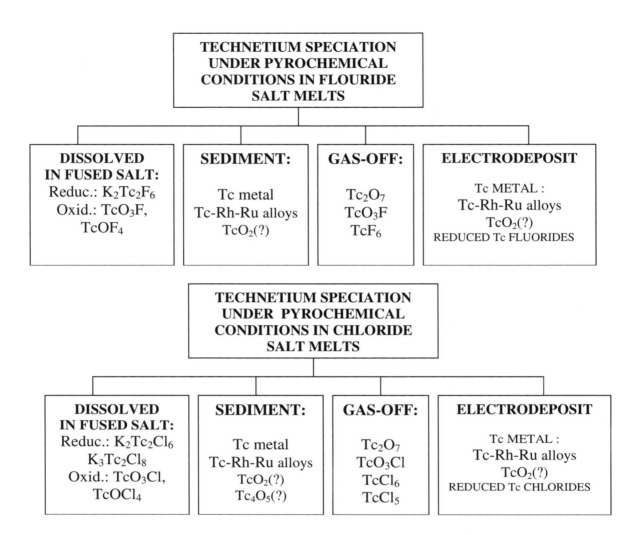

Figure 1. Radiochemical hutch of ROBL/ESRF

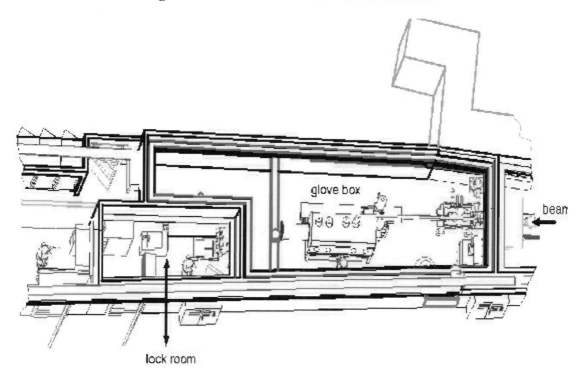

Figure 2. NMR ^{99}Tc spectrum of Tc metal powder (ϕ80-150 µm)

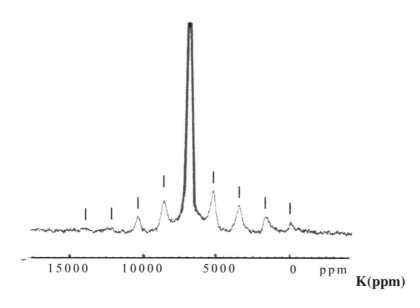

Figure 3. Temperature dependence of isotropic Knight shift (K) for Tc metal powder

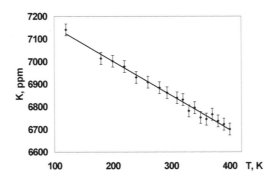

Figure 4. Temperature dependence of ν_Q for Tc metal powder

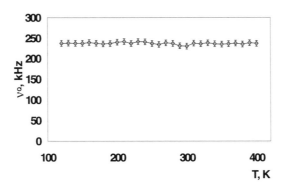

Figure 5. ^{99}Tc NMR spectrum of Tc foil (d = 20 μm)

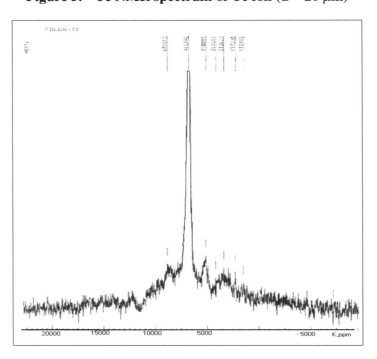

Figure 6. Tc K-edge k^3-weighted EXAFS and corresponding Fourier transform of sample (Me₄N)₂TcBr₆

EXAFS structural parameters

Tc-Br	N = 5.8(2)	R = 2.51(2) Å	σ^2 = 0.0040 Å^2	ΔE_0 = -16.9(5) eV

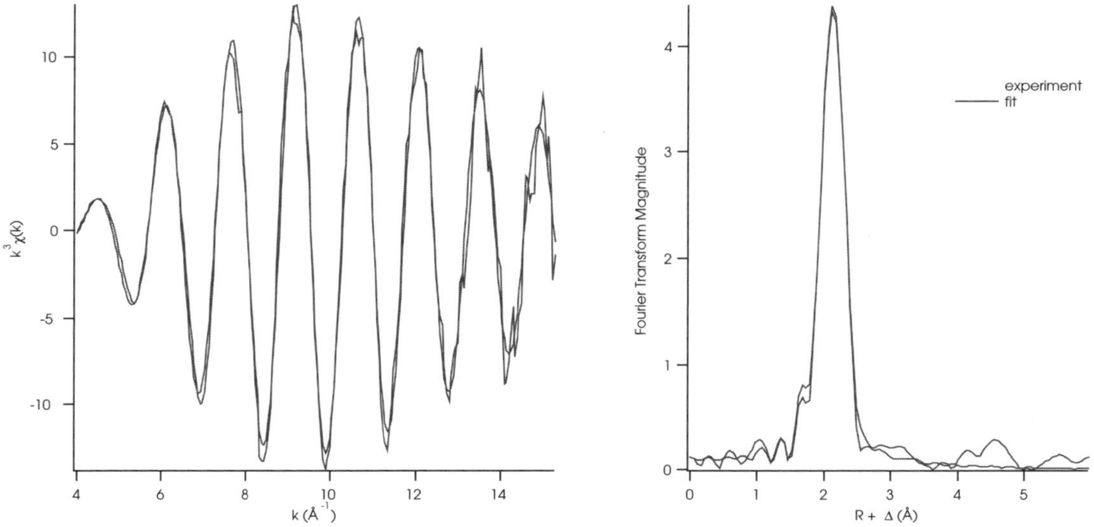

Figure 7. Tc K-edge k^3-weighted EXAFS and corresponding Fourier transform of sample K₃Tc₂Cl₈

EXAFS structural parameters for K₃Tc₂Cl₈ (preliminary approximation)

Tc-Tc	N = 1.6(3)	R = 2.20(2) Å	σ^2 = 0.0069 Å^2	ΔE_0 = -1.1(9) eV
Tc-Cl	N = 2.2(4)	R = 2.46(2) Å	σ^2 = 0.0107 Å^2	

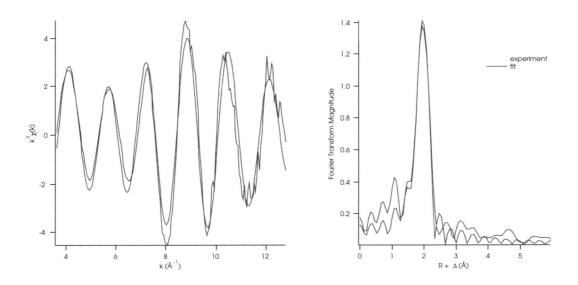

MOLTEN SALT BURNER FUEL BEHAVIOUR AND TREATMENT

Victor V. Ignatiev, Raul Ya. Zakirov
RRC-Kurchatov Institute
Kurchatov sq.1, Moscow, 123182, Russian Federation

Konstantin F. Grebenkine
RFNC – All-Russian Institute of Technical Physics
Snezhinsk 454070, Russian Federation

Abstract

The objective of this paper is to discuss the feasibility of molten salt reactor technology for treatment of Pu, minor actinides and fission products, when the reactor and fission product clean-up unit are planned as an integral system. This contribution summarises the available R&D which led to selection of the fuel compositions for the molten salt reactor of the TRU burner type (MSB). Special characteristics of behaviour of TRUs and fission products during power operation of MSB concepts are presented. The present paper briefly reviews the processing developments underlying the prior molten salt reactor programmes and relates them to the separation requirements of the MSB concept, including the permissible range of processing cycle times and removal times. Status and development needs in the thermodynamic properties of fluorides, fission product clean-up methods and container materials compatibility with the working fluids for the fission product clean-up unit are discussed.

Introduction

The use of the molten salts as a fuel material has been proposed for many different reactor types and applications. Particularly in the US, Russia, France and Japan molten salt fuelled reactor (MSR) concepts have been developed for fast breeders and thermal reactors. Though the molten salt nuclear fuel concept has been proven successful through the operation experience of a MSRE experimental reactor at ORNL [1,2], this approach has not been implemented in industry. A mixture of ^{7}LiF-BeF$_2$-ZrF$_4$(-ThF$_4$)-UF$_4$ was the fuel chosen for operation of MSRE and for subsequent reactors of this type.

In Russia the MSR programme began in the second half of 70s. RRC-Kurchatov Institute was a basic organisation under whose supervision a collaboration of specialised institutions was formed and functioned. A reduction of activity occurred after 1986 due to the Chernobyl accident, followed by a general stagnation of nuclear power and nuclear industry. Then, at the end of 80s, there was an increase of conceptual studies as a result of the interest in inherently safe reactors of new generation. An extensive review of MSR technology developments at RRC-KI through 1989 is given in Ref. [3].

The interest in exploring molten salts as a future option in nuclear power both for the back end of the fuel cycle or for its overall simplification is currently being revisited [4-9]. Particularly, chloride salts have also been proposed as an alternative fluid fuel to obtain a fast neutron spectrum due to high solubility of Pu and transuranics (TRU) in the melt, but severe problems related to structural materials corrosion, chemical stability of such systems and poor separation ability between some representatives of actinides and lanthanides groups have been pointed out [9,10]. Also, during irradiation ^{35}Cl transmuted to ^{36}Cl with $T_{1/2} = 300\ 000$ yr. Ternary and quaternary systems of fluoride fuels still remain an interesting way out [11,12]. Desirable characteristics of these innovative fuels are:

- Stability and negligible pressurisation at high temperatures.

- Good hydrodynamic and heat transfer properties.

- Stability in high radiation field.

- Low neutron absorption cross-sections.

- Low solubilities of gaseous fission products, which make possible a several fold increase in burn-up time in the system.

Between 1994-1996, a feasibility study of molten fluoride systems as applied to Pu burning and long-lived radwaste transmutation in accelerator-driven system (ADS) had been supported by the International Science and Technology Centre (ISTC) under the Project #17 [8]. As a continuation of this project, Russian institutes have submitted to the ISTC the new Project #1606, which would undertake the experimental study of molten salt technology for safe, low waste and proliferation resistant treatment of radwaste and plutonium in accelerator-driven and critical systems. Currently some study on the feasibility of a new concept of MSB fuelled by Pu and minor actinides is being supported by MINATOM.

The new MSB concept requires a reconsideration of the prior concept of MSRs, including optimisation of the neutron spectra in the core and R&D which led to a selection of the salt composition and approaches to its clean-up. The present paper is concerned with the MSB concept as regards the choice of the fuel composition, fission product clean-up methods and structural materials for fission product clean-up facility.

Nuclear design and fuel cycles flexibility

The introduction of the innovative reactor concept of the incinerator type in the future nuclear power system should provide:

- Low plutonium and minor actinides total inventory in the nuclear fuel cycle (M).

- Reduced actinides total losses to waste (W).

- Minimal ^{235}U support.

- Minimal neutron captures outside actinides (coolant and structural material activation products).

Estimations have shown a strong dependence of the first two parameters (M and W), which are responsible for incinerator efficiency, from the burn-up (c) reached in core of incinerator and actinides mass flow rate in the fuel cycle (A(t) = G(t)/Q(t), where G(t) is the amount of TRU fed to the process during t, and Q(t) is the electricity produced during t). For example, in multi-recycling mode and co-processing case with the assumption that single cycle losses in fuel reprocessing (a) and fabrication (b) -a,b << c we could obtain the following approximate equations:

$$M \approx A(\tau) \cdot \tau \cdot \left[1 + (a+b+c)^{-1} \right] \approx A(\tau) \cdot \tau \cdot (1+c)/c , \text{ kg TRU/GWe (at equilibrium)} \qquad (1)$$

$$W = W_{rec} + W'_{rem} \approx A(t) \cdot \left[(a+b)/c + (1-c)^{t/\tau} \right], \text{ kg TRU/TWhe (at the end of scenario)} \qquad (2)$$

where W_{rec} is recycling losses, W'_{rem} is residual inventory, and τ is the turnaround time of one recycle.

The interest in the MSR of incinerator type stems mainly from an increased burn-up time in the system, reduced actinides mass flow rate and relatively low waste stream when purifying and reconstituting the fuel. It is natural to expect that in the future molten salt reactor (MSR) technology could find a role in symbiosis with standard reactors in the management of plutonium, minor actinides and possibly fission products.

The advantages of the MSR as a burner reactor follow not only from a possibility for its effective combination with the dry technique of fuel processing, which has the prospect of being of low cost and producing a small volumes of wastes, but also from its capability to use fuel of different nuclide compositions. The MSRs have the flexibility to utilise different fissile fuel in continuous operation with no particular modification of the core, as it was demonstrated during MSRE operation for $^{233,235}U$ and Pu. Further, the MSRs can tolerate denaturing and dilution of the fuel, as well as contamination by lanthanides [1].

The results of recent studies have demonstrated that a broad range of MSBs operating in critical or accelerator-driven modes with PuF_3 and minor actinides as the fuel is conceptually feasible [4-7]. At one end of this range is the well-thermalised graphite moderator reactor, in which fluid fuel consists of a molten mixture of $^7Li,Be/F$ or $Na,Be/F$ containing appropriate quantities plutonium and minor actinides as trifluorides. At the other end is a MSB operating with fast spectrum without graphite moderator, in which the solvent system prepared from $NaF-^7LiF$ (and/or other possible constituents, like ZrF_4, PbF_2, CaF_2). This MSB concept could have a concentration of PuF_3 much higher than that of previous one. Phase behaviour of some such mixtures appears suitable to permit the use of a high concentration of PuF_3 in melts whose freezing point will be acceptable for single fluid MSBs. The basic reactor flowsheet for the MSB is essentially the same as that for the reference ORNL designs [1,2], in which a single molten salt containing both fissile and fertile materials serves as both the fuel and coolant. The only differences are in the core/blanket configuration, details of the fuel salt composition and the fission product clean-up system. U-free fuel matrix, as well as the addition of ThF_4 and UF_4 in homogeneous solution, is conceptually feasible for MSB.

In MSR the molten salt is pumped from the core to heat exchanges where heat generated by fission is transferred to a molten secondary salt, a eutectic mixture of $NaBF_4$-NaF. The secondary salt transports the heat to the steam supply system and in the case of LiF constituent use, also serves to intercept tritium migrating through the heat exchange system toward the steam circuit.

In MSRs all the fission product species do not go to the processing plant; Kr and Xe are removed by sparging with helium (50 sec) in the reactor. The semi-noble and noble metals rapidly transport to purge gas and deposit on surfaces (2.5 hours) within the fuel circuit, where most of it is retained [1]. For MSR one of the main functions of the processing unit is to remove soluble rare earth fission products. It is clear that the processing flowsheets in MSB must differ from prior MSR designs in some important aspects. Possible conceptual lines are listed below (see Figure 1).

The simpler of the MSB concepts would completely eliminate on-line chemical processing of the fuel salt for removal of fission products. In order to achieve high burn-up results the stripping of gaseous fission products would be retained, and some batch-wise treatment to control oxide contamination probably would be required. This reactor needs routine additions of TRU fuel, probably with some ^{235}U support, but would not require replacement or removal of the in-plant inventory except at the end of the plant life time (t_p). If we denote actinides concentration in the fuel by ρ_c, heat generated in actinides by B, plant efficiency by η, capacity factor by φ and core/auxiliaries volumes by V_{in}/V_{out} we can calculate actinide mass flow rate $A(t_p)$ for one reactor lifetime from the following equation:

$$A(t_p) \approx \left[\rho_c \cdot (1 + V_{out}/V_{in})/(q_c \cdot t_p) + c/B \right]/(\eta \cdot \varphi), \text{ kg TRU/TWhe} \qquad (3)$$

The primary feature in the MSB concepts is a high-power density (q_c). The maximum power density in the core is limited by fast neutron damage to graphite moderator/reflector. Indeed the principle shortcoming of the single fluid concept is the substantial investments of the fissile material in the heat transfer circuit and elsewhere outside the reactor core (V_{out}). Obviously, these investments will increase with the increase of the core specific power q_c, because the removal power density in the external power circuit is limited primarily by heat transfer and pressure drop consideration. An optimum salt discard rate exists, for which the fuel burn-up time is balanced against the increasing of the core inventory and fuel make-up, which increases due to neutron balance worsening. In addition, there is a minimum discard rate required to limit the concentration of An/Ln trifluorides in the circulating fuel to an acceptable level (see the following section).

Adding a batch or on-line chemical facility to the single-pass reactor provides the other conceptual line, based on uranium and TRU recycling, as often as necessary. The fluoride volatility method, converting UF_4 to gaseous UF_6 and uranium oxide precipitation are proven methods for recovery of uranium from molten fluoride salt. Since plutonium and minor actinides must be removed from the fuel solvent before rare earth fission products, the MSB must contain a system that provides for removal of TRUs from the fuel salt and their reintroduction to the fresh or purified solvent (see Figure 1). This plutonium reintroduction circuit has the advantage of also returning americium, curium and californium to the reactor fuel and permitting only very small losses of any TRUs to the waste stream. Since the higher actinides would always accompany the plutonium, this operation would never produce a "clean" material that would be attractive for diversion. Quite satisfactory MSB characteristics could be achieved in the absence of the salt processing for fission products, with the occasional batch discard and replacement of the entire inventory of carrier salt, not including TRUs. With the addition of the extractive processing for soluble fission products, such a reactor could reuse the purified solvent (see Figure 1). Certainly, the processing cycle time in the MSB for these species can be considerably greater than that for TRUs (e.g. $2\tau_{An}, \approx \tau_{salt}$, where τ_{An}, τ_{Ln} are the turnaround times of one recycle for An and purified solvent).

There is a need for determination by computer in close co-ordination with MSB neutronic studies, the permissible range (and the optimum) of processing rates and modes as well as salt replacement interval, which we believe will result in minimal TRU mass flow rate and maximum burn-up.

The fuel salt for MSB concept

Many chemical compounds can be prepared from several "major constituents". Most of these, however, can be eliminated after elementary consideration of the fuel requirements. Consideration of nuclear properties alone leads one to prefer as diluents the fluorides of Be, Bi, ^{7}Li, Pb, Zr, Na, and Ca, in that order. The simple consideration of the stability of these fluorides toward reduction by structural metals, however, eliminates the bismuth fluorides from consideration (see Table 1).

Note that ZrF_4, as a part of basic solvent, was found to distil from melt and condense on cooler surfaces in the containment system [1]. Control of the ZrF_4 mass transport was considered too difficult to ensure, so the $2LiF-BeF_2$ solvent system was chosen at ORNL. Also, in order to minimise problems associated with chemical treatment of the fuel salt and associated reduction of the basic components, priority should be given to the system with lowest possible ZrF_4 and PbF_2 content. Note that use of Zr and Pb, instead of, e.g. the sodium, in the basic solvent will lead to the increased generation of the long-lived activation products in the system.

Trivalent plutonium and minor actinides are the only stable species in the various molten fluoride salts [11,14]. Tetravalent plutonium could transiently exist if the salt redox potential was high enough. But for practical purposes (stability of potential container material) salt redox potential should be low enough and corresponds to the stability area of Pu(III).

PuF_3 solubility is maximum in pure LiF or NaF and decreases with the addition of BeF_2 and ThF_4. The decrease is more for BeF_2 addition, because the PuF_3 is not soluble in pure BeF_2. The solubility of PuF_3 in $LiF-BeF_2$ and $NaF-BeF_2$ solvents is temperature and composition dependent and PuF_3 solubility seems to be minimal in the "neutral" melts. For the latter it reachs about 0.5 mol% at 600°C and increases to about 2.0 mol% at 800°C. For some excess free fluoride solvent system Li,Be,Th/F (72-16-12 mol%) studies indicated that solubility of PuF_3 increased from 0.7-0.8 mol% at 510°C to 2.7-2.9 mol% at 700°C.

The other TRU species are known to dissolve in Li,Be,Th/F solvent, but no quantitative definition of their solubility behaviour exists. Such definition must of course be obtained, but the generally close similarity in behaviour of the AnF_3 makes it most unlikely that the solubility of this individual species could be a problem. As expected, substitution of a small quantity of AnF_3 scarcely changes the phase behaviour of the solvent system.

The trifluoride species of AnF_3 and the rare earths are known to form solid solutions so, that in effect, all the LnF_3 and AnF_3 act essentially as a single element. It is possible, but highly unlikely, that the combination of all trifluorides might exceed this combined solubility at a temperature below the reactor inlet temperature. A few experiments must be performed to check this slight possibility. The solubilities of the AnO_2 in Li,Be,Th/F are low and well understood. Plutonium as PuF_3 shows little tendency to precipitate as oxide even in the presence of excess BeO and ThO_2. The solubility of the oxides of Np, Am, Cm has not been examined. Some attention to this problem will be required.

As expected, the single fluid fast spectrum MSB will need a concentration of PuF_3 much higher than that for the $2LiF-BeF_2$ system with graphite moderator. The $LiF-PuF_3$ system has eutectic at 743°C and 20 mol% PuF_3. The $NaF-PuF_3$ system has eutectic at 24 mol% PuF_3, melting at 727°C [11]. Inspection of the diagram for Li,Na,Pu/F system reveals that a considerable range of compositions with about 10 mol% PuF_3 will be completely molten at 600°C. It is possible that some addition of e.g. CaF_2 would provide liquids for quaternary system at lower temperatures. As expected from the general similarity of PuF_3 and minor actinides trifluorides, substitution of a relatively small quantity of AmF_3, CmF_3, NpF_3 for PuF_3 scarcely changers the phase behaviour. Accordingly, the phase behaviour of the fuel will be dictated by that of the Li,Na,Pu/F system. Significant quantities of minor actinides in the mixture will complicate phase behaviour of the fuel.

The molten fluoride chemistry (solubility, redox chemistry, chemical activity, etc) for the $2LiF-BeF_2$ system is well established and can be applied with great confidence, if PuF_3 fuels are to be used in the $2LiF-BeF_2$ solvent. But the chemistry of other solvent systems are different and less understood (for example, in the more acidic Li,Na/F system) and requires more comprehensive study. For MSB needs, another important consideration is that of PuF_3 chemical behaviour in these solvent systems: PuF_3 solubility in Li,Be/F and Na,Be/F solvents; phase transition behaviour of the ternary Na,Li,Pu/F and quaternary Na,Li,Ca,Pu/F fuels, oxide tolerances of such mixtures and redox effects of the fission products.

Fission product clean-up

For molten salt fuels, fission products could be grouped by the three broad classes: 1) the soluble at salt redox potential fission products, 2) the noble metals and 3) the noble gases. The MSB would manage the noble gas removal by sparging with helium in the same manner as other MSR concepts [1,2]. The problem here is to prevent the xenon from entering the porous graphite moderator. Development of sealing techniques should be continuing. For the noble metals the situation is not so good, and more experimental efforts are required in order to control their agglomeration, adhesion to surfaces and transport in purge gas.

In MSBs, from which xenon and krypton are effectively removed, the most important fission product poisons are among lanthanides, which are soluble in the fuel. A number of pyrochemical processes (reductive extraction, electrochemical deposition, precipitation by oxidation and their combinations) for removing the soluble fission products from the fluoride-based salt have been explored in recent years.

Important parameters for chemical treatment are free energy of formation $\Delta G_f^0(T)$ of substances and separation factors of elements between molten salt and liquid metal θ [14]. These separation factors for different elements with the same valence (e.g. actinides and lanthanides) could be given by:

$$\ln \theta = \left\{\Delta G_f^0(Ln, T) - \Delta G_f^0(An, T)\right\}/(RT) \tag{4}$$

where T is the temperature (K), R is the molar gas constant ($J \cdot mole^{-1} \cdot K^{-1}$).

Preliminary thermodynamic consideration predicts the need for greater efforts concerning the separation of Am and Cm from Ce and La in a chloride system in comparison with a fluoride one (Figure 2). Note that available thermodynamic data (calculated or measured) for An/Ln trifluorides are taken from different sources and mostly refer to pure compounds rather than to species in solutions (see Table 1). They include considerable uncertainties and dispersion that does not permit their use while determining An/Ln separation performances. There is a need for experimental study on An/Ln trifluorides thermodynamic properties using unified methods (for example, electrochemical method using solid-electrolyte cells).

Eq. (4) shows that the separation factors strongly depend on the difference between $\Delta G_f^0(Ln, T)$ and $\Delta G_f^0(An, T)$. It is important that for An/Ln trifluorides, due to their chemical similarity difference between $\Delta G_f^0(Ln, T)$ and $\Delta G_f^0(An, T)$, for each of the species involved in each equilibrium with and without solvents to remain practically constant.

Previous studies have shown the significant impact of metallic solvents on thermodynamic constants and activity coefficients. Deviations may be very large. Estimations showed that the available options for metallic solvents in the application of pyrochemistry to the MSB fission product clean-up are relatively limited (e.g. Bi, Cd for recycling and Al, Zn for single pass approach). Of the processes mentioned above for fluoride systems, reductive extraction into bismuth is being most developed at ORNL [1]. The Bi-Li alloy is used for reductive extraction from different melts according to the scheme:

$$MF_n(\text{melt}) + nLi^0(\text{Bi}) = M^0(\text{Bi}) + nLiF(\text{melt}) \qquad (5)$$

In this process the rare earths are extracted from a fuel salt stream into lithiated bismuth after all the plutonium and minor actinides have been previously removed. The separation factors θ for Li,Be/F and Li,Na,K/F solvent systems are approximately equal to 10^3 ($\theta = (X_{An}/X_{AnF_n})(X_{LnF_n}/X_{Ln})$, where X_{An}, X_{Ln} are the mole fractions of An and Ln in alloy, X_{AnF_n}, X_{LnF_n} are the mole fraction of AnF_n and LnF_n in a molten salt). These values are very convenient for the lanthanide separation. However, when the melt is complicated by the addition of large quantities of ThF_4 the situation becomes considerably less favorable ($\theta_{Ln-Th}\sim 1$). The rare earth removal unit based on Bi-Li reductive extraction flowsheet developed in ORNL for a LiF-BeF$_2$ solvent system could provide negligible losses of TRU ($\approx 10^{-4}$) through the use of several counter current stages.

For fluoride based systems alongside with reductive extraction other metal transfer methods of lanthanide removal could be considered. Particularly, electrochemical deposition of elements as metals on liquid or solid electrodes in electrochemical cells, filled with fuel salt and bismuth or lead. The applicability of a given method for fluoride based systems is defined by: 1) the fuel salt being at a temperature of 500-800°C producing a good electrolyte, the conductivity of which is more than $1 \text{ om}^{-1}\text{cm}^{-1}$; 2) in LiF–NaF salt solution the standard potentials for salt components (LiF, NaF) are higher than that for actinide and lanthanide trifluorides. Certainly, the choice of electrode type (solid vs. liquid) depends on the process requirements.

Although the metal-transfer process appears to give the best fuel salt purification in the case of a processing system with a relatively short cycle (10-30 days), there are other possibilities for rare earth removal that are perhaps worth keeping in mind. If the bismuth containing system proves expensive or if unseen engineering difficulties develop (e.g. material required), other methods may be applicable, especially at longer processing cycle times. If treatment of a MSB fuel on a cycle time of 100 days or more is practicable, such an oxide precipitation might be used for periodic removal of rare earths.

In experiments [15] a successful attempt was made to precipitate mixed uranium, plutonium, minor actinides and rare earths from a LiF-NaF molten salt solution at 700-900°C by $CaO(Al_2O_3)$ oxidation. The rare earth concentration in the molten salt solution was about 5-10 mol%. The following order of precipitation in the system U-TRU-Al-Ln-Ca was found. Essentially all U and TRU were recovered from the molten salt till to rest concentration $5 \cdot 10^{-4}\%$, with rare earths still concentrated in solution. For U/Th fuelled salts this suggests that, after the Pa and U have been precipitated as oxides, the thorium may also be separated out in this way, leaving the carrier salt containing TRU and rare earths for the following processing steps. A drawback of oxidation and electrochemical deposition is that they are intrinsically of batch type, and so likely to be more expensive on a large scale than that of the continuous mode preferred for industrial purposes.

Several combinations of the above processes with some of the alternatives are being considered for use with the MSB rare earth removal unit. With these and other possibilities for chemical processing relatively unexplored, there seems to be much room left for interesting studies in the units when TRU are the fuel.

Materials for the fission product clean-up unit

The materials required for a fission product clean-up unit depend of course, upon the nature of the chosen process and upon the design of the equipment to implement the process. For MSB the key operation in the fuel treatment is removal of rare earth, alkali metal and alkaline earth fission products from the fuel solvent before its return, along with the TRUs, to the reactor. The crucial process in most of the processing vessels is that liquids be conducted to transfer selected materials from one stream to another.

Such a fission product clean-up unit based on metal transfer process at least will present a variety of corrosive environments, including:

- Molten salts and molten alloys containing e.g. bismuth, lithium or other metals at 650°C.

- HF-H_2 mixtures and molten fluorides, along with bismuth in some cases, at 550-650°C.

- Interstitial impurities on the outside of the system at a temperature of 650°C, particularly if graphite and refractory metals are used.

Certainly, the R&D on the materials for the fission product clean-up unit for MSB is at a very early stage. A layer of frozen salt will probably serve to protect surfaces that are worked under oxidising conditions. If the layer can be maintained in the complex equipment, RRC-KI preliminary tests at molten salt loops [16] showed that the thickness of the frozen film on the wall was predictable and adhered to the wall.

The only materials that are truly resistant to bismuth and molten salts are refractory metals (W, Mo, Ta) [1] and graphite (e.g. graphite with isotropic pyloric coating tested in RRC-KI [16] for both working fluids), neither of which is attractive for fabricating a large and complicated system [5,6]. Development work to determine if iron base alloys can be protected with refractory metal coatings should probably be considered for higher priority. The approach taken to materials development could be to initially emphasise definition of the basic material capabilities with working fluids and interstitial impurities, and then to develop a knowledge of fabrication capabilities.

Summary

Though the molten salt nuclear fuel concept has a solid background, its specific application for the management of plutonium and minor actinides requires additional study. The major uncertainties are in the areas of TRU trifluorides behaviour, fuel salt processing and materials for the fission product clean-up unit.

To solve some of the essential issues mentioned in previous sections, Russian institutes (RFNC – All-Russian Institute of Technical Physics, (Chelyabinsk-70), RRC-Kurchatov Institute (Moscow), Institute of Chemical Technology (Moscow) and Institute of High Temperature Electrochemistry (Ekaterinburg)) have submitted to the ISTC the Project #1606, concerning experimental study of molten salt technology for safe, low waste and proliferation resistant treatment of radwaste and plutonium in accelerator-driven and critical systems.

The major developments that we believe should be pursed in the framework of the project are the following:

- MSB conceptual designing and development.

- Experimental study of behaviour and fundamental properties for prospective molten salt fuels.

- Experimental verification of candidate structural materials for fuel/coolant circuits and fission product clean-up unit.

The last two objectives are considered crucial to the MSB concept further development. The experimental data would be fed into the conceptual design efforts. The objectives of the conceptual design and development programme are to first identify candidate flowsheets (reactor + fission product clean-up unit) for the MSB concept, that will be technologically feasible. Then, based on experimental data received in the project, the choice of optimal flowsheet, including fuel composition and design parameters, will be made.

One of the aims of this contribution is to attract the attention of potential foreign partners with regard to the ISTC Project #1606, and to invite them to take part in collaboration under the terms of the project.

REFERENCES

[1]   H.J. MacPherson, *Reactor Technology*, Vol. 15, N. 2, p. 136 (1972).

[2]   J.R. Engel, *et al.*, ORNL/TM-6415, March (1979).

[3]   V.V. Ignatiev, *et al.*, "Molten Salt Nuclear Energy Systems", Energoatomizdat, Moscow (1990).

[4]   V.V. Ignatiev, K.F. Grebenkine, R.Y. Zakirov, "Experimental Study of Molten Salt Reactor Technology for Safe, Low-Waste and Proliferation Resistant Treatment of Radioactive Waste and Plutonium in Accelerator-Driven and Critical Systems", in Proc. of GLOBAL'99, Jackson Hole, USA (1999).

[5]   J. Vergnes, *et al.*, "Limiting Plutonium and Minor Actinides Inventory: Comparison Between Accelerator-Driven System (ADS) and Critical Reactor", ibid.

[6]   C. Bowman, "Accelerator-Driven Transmutation of Waste Using Thermal Neutrons: High Burn-up and Weapons Material Elimination Without Recycling", ibid.

[7]   I.S. Slessarev, *et al.*, Concept of the Thorium Fueled Accelerator-Driven Subcritical System for both Energy Production and TRU Incineration", in Proc. of ADTTA'99, Praha, Czech Rep. (1999).

[8]   K.F. Grebenkine, in Proc. of ADTTA'96, Kalmar, Sweden, p. 221 (1996).

[9]   V.V. Ignatiev, *et al.*, in Proc. of the Fifth Information Exchange Meeting on Actinide and Fission Product Partitioning and Transmutation, Mol, Belgium, p. 169 (1998).

[10]   L.L. Migai, T.A. Taritsina, "Corrosion Resistance of Structural Materials in Halides and Their Mixtures", Metallurgy, Moscow (1988).

[11]   C.J. Barton, J.D. Redman, R.A. Strelow, *J.Inorg. Chem.*, Vol .20, N. 1-2, p. 45 (1961).

[12]   S. Cantor, ORNL-TM-2316 (1963).

[13]   V.F. Afonichkin, *et al.*, in Proc. of ADTTA'96, Kalmar, Sweden, 3 June, Vol. 2, p. 1144 (1996).

[14]   V.A. Lebedev, *J. Melts*, Vol. 10, p. 91 (1997).

[15]   V.F. Gorbunov, *Radiochimija*, Vol. 17, p. 109 (1976).

[16]   V.V. Ignatiev, *et al.*, Preprint IAE–5678/11, Moscow (1993).

Table 1. Thermodynamic properties of fluorides

Compound (solid state)	$-\Delta G_{f,1000}$ kcal/mol	$-\Delta G_{f,298}$ kcal/mol	E_{298}, V (Me|F$_2$)
LiF	125	140	6.06
CaF$_2$	253	278	6.03
NaF	112	130	5.60
BeF$_2$	208	231	5.00
ZrF$_4$	376	432	4.70
PbF$_2$	124	148	3.20
BiF$_3$	159	200	2.85
NiF$_2$	123	147	3.20
UF$_3$ (UF$_4$)	300 (380)	330 (430)	4.75
PuF$_3$ (PuF$_4$)	320	360 (400)	5.20
ThF$_4$	428	465	5,05
AmF$_3$	325	365	5.30
CeF$_3$	345	386	5,58
LaF$_3$	348	389	5,63

Figure 1. Possible processing flowsheets for MSB concepts

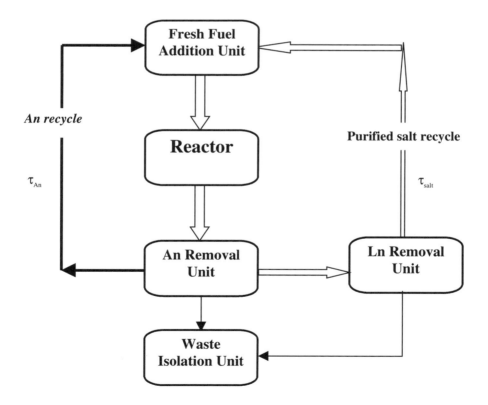

Figure 2. Separation factors between actinides and lanthanides for chloride (T = 800 K) and fluoride (T = 1 000 K) systems

Fluorides

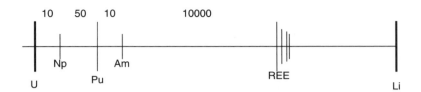

Clorides

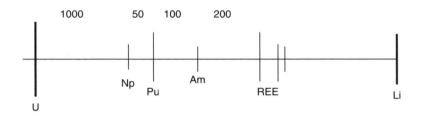

RECOVERY OF NOBLE METALS (RU, RH, PD, MO) AND TELLURIUM FROM FISSION PRODUCTS

C. Jouault, M. Allibert
LTPCM/INPG, 1130 rue de la piscine
B.P. 75, Domaine Universitaire
38402 St Martin d'Hères Cedex France
E-mail: cjouault@ltpcm.inpg.fr

R. Boen, X. Deschanels
DCC-RRV-SCD, CEA/Valrhô-Marcoule
BP 171, 30207 Bagnols-sur-Cèze, France

Abstract

Prior to pyrochemical separation actinides/lanthanides, caesium, noble metals and other easily reducible species must be extracted from fission product mixed oxides. Tellurium oxide, as well as caesium oxide, are first extracted in the course of a processing by a reducing gas (CO or H_2); they volatilise when they are metallic. After a processing by gaseous fluorhydric acid, the separation of the other species (Ru, Rh, Pd, Mo…), which have also been reduced by CO or H_2, is accomplished by the settling of these metallic elements from the fluoride salt (LiF-CaF_2-AcF_x-LnF_y) to a liquid metal (Sb, Sn, Zn). This process is known as "digestion". The digestion of metallic particles by a liquid metal seems to be controlled by problems of wettability between these particles, the molten fluoride salt and the digesting metal.

Introduction

The first axis of the 1991 French law concerning the treatment of nuclear waste deals with the separation and transmutation of the fission products. Three main groups of products must be separated, depending on their chemical properties: noble metals and other easily reducible species (Te, Sn, Fe, Sb, Ni,...), actinides and lanthanides. One way to achieve this separation is to use pyrochemical techniques, as for example with liquid-liquid extraction. This technique consists of bringing into contact a molten salt containing the fission products and an adequate liquid metal. The latter reduces a group of elements having the same thermodynamic potential to the metallic state so as to dissolve them.

Actinide/lanthanide separation has been widely studied in chloride or fluoride salts. However, the first step of this process of separation of fission products, that is the noble metals/actinides separation, has not been closely examined. The study we led deals with this separation in fluoride salt.

In the literature, studies dealt with the pyrochemical separation of noble metals (Ru, Rh, Pd, Mo) as oxides [1] or metallic alloys [2,3] from an oxide salt or as chlorides [4] from a chloride salt. The liquid metals were typically Pb, Bi, Sb, Cu, Sn, Zn, Cd. The authors found that Pd and Rh were well reduced (if necessary) and recovered in the metallic phase whatever the experimental conditions. On the contrary, the recovery of Ru and Mo depends on the conditions and, in any case, was incomplete. These elements were also well reduced, but not completely separated from the molten salt. They were believed to remain as metallic particles between the salt and the metallic phase.

Process of separation of the fission products

A global process of separation of the fission products was elaborated from thermodynamic calculations and bibliographical researches. This process is based on the differences in thermodynamic potentials existing between the various elements (Figure 1). The first elements to extract are noble metals (Mo, Ru, Rh, Pd) and the easily reducible species (Sn, Fe, Ni, …).

F. Lemort [5] has studied the pyrochemical actinides/lanthanides separation in a fluoride salt. He assumed the fluoride flux that entered in his process was free of noble metals. Fluoride salt is chosen because it will facilitate the final vitrification of the waste.

Initially, the flux of fission products is made of oxides. Oxides of noble metals and tellurium can not be transformed into fluorides by gaseous fluorhydric acid as actinides or lanthanides could. It was thus impossible to extract them using the same process used by F. Lemort (reduction by an adequate metal of the wanted fluorides and dissolution of the formed metals into the metallic phase). Therefore, the idea was to reduce them to the metallic state using a gas at the beginning of the process when the fission products are pulverulent, and then to extract them by dissolution into a metallic phase. This has the advantage that no reducing agent is added to the flux.

The process (Figure 2) begins with a treatment of the mixed oxides by a reducing gas, CO or H_2. The easily reducible species are reduced to the metallic state and among them, caesium and tellurium volatilise. After a treatment by gaseous fluorhydric acid (to transform the oxides of actinides and lanthanides into fluorides) and dissolution into a fluoride solvent, the mixture of metallic species (noble metals, Ni, Sn, Sb...) and fluorides are brought into contact with a liquid metal which is supposed to digest the metallic particles.

The reduction of tellurium oxide by CO or H_2 and the volatilisation of tellurium were confirmed by experimental results. The digestion of the metallic species by a liquid metal has also been experimentally studied.

Tellurium oxide extraction: Experimental results

Thermodynamically, the oxides of noble metals, caesium, tellurium and other easily reducible species can be reduced by H_2 or CO. The oxides of actinides or of lanthanides can not be reduced by these gases.

In the literature, confirmation exists of the reduction for all the elements except for tellurium. We just learned that H_2 reduces tellurium oxide only partially. No study on reduction by CO was found.

In order to clarify this first step of processing of fission products, we rapidly studied the reduction of tellurium oxide by H_2 or by CO. The results for the reduction by CO (100 ml/min) can be seen in Figure 3. In fact, we never saw tellurium metal. We just noticed a loss of substances. This loss could have two origins: the volatilisation of TeO_2 or the volatilisation of Te, as soon as it is obtained. In conclusion, experiments were carried out with argon instead of hydrogen, and no loss was observed. We can infer that CO reduces TeO_2 to Te that volatilises thanks to the flux of gas. We never obtained a complete loss. The limiting reaction is the reduction since at the end of the experiment, the substance observed in the crucible was TeO_2 and not Te. The two main parameters seem to be the state of TeO_2 and the temperature; the solid state facilitates the reaction by a higher surface of contact. In a given state, TeO_2 is more rapidly reduced at a higher temperature.

Two experiments were carried out under H_2 (200 ml/min), one with solid TeO_2 during 5 h and one with liquid TeO_2 during 8 h. In each case, TeO_2 was completely reduced to Te that volatilised. Thus, the reduction by H_2 is kinetically better than the reduction by CO.

Consequently, the first step of reprocessing of fission products may be a reduction of the oxides of noble metals and of easily reducible species by a gas.

Noble metals recovery: Case of molybdenum and ruthenium

As we saw before, noble metals and other easily reducible species are extracted from the molten fluoride salt by digestion in a liquid metal.

This process of digestion is not peculiar to our study: in the study of pyrochemical actinides/ lanthanides separation (chloride or fluoride salt), Moriyama *et al.* [6-8] undertook to solve a similar phenomenon. Indeed, they observed that actinides/lanthanides separation, once the actinides have been reduced, consists in fact of a digestion in the liquid metal of the metallic or inter-metallic particles formed in the salt.

Experimental

The criterions for choosing the digesting metals were mainly their melting point, their potential of reduction (they should not reduce the actinide fluorides) and the solubility of the various elements to digest or, the possibility of making inter-metallic compounds. The search for these characteristics led to the choice of antimony, tin and zinc.

The first elements for which digestion was studied were molybdenum (3-4 μm) and ruthenium (75 μm) because they are the least soluble of all the metals that it is possible to use here, and so they are supposed to be the most difficult to digest into the liquid metal.

The metal was either cast into the pre-melted fluoride salt (LiF$_{(49.5\ wt.\%)}$-CaF$_{2(38.5\ wt.\%)}$-Mo$_{(12\ wt.\%)}$ or LiF$_{(55.2\ wt.\%)}$-CaF$_{2(42.8\ wt.\%)}$-Ru$_{(2\ wt.\%)}$) or was put under the salt from the beginning. The experiment lasted two hours at 790°C with a stirring rate of 40 rpm under argon or hydrogen. The final concentration of Mo or Ru in the fluoride salt was determined by inductive coupled plasma spectrometry. The reactor in which the experiments were carried out is shown in Figure 4.

Results and discussion

Results under argon

The first experiments were conducted under argon. Of all the liquid metals, Ru had the best recovery rate; 95% of Ru was in the metal. Further experiments must be performed with smaller particles. For Mo, Sn did not lead to a good recovery – 20 wt.% of Mo was extracted – whereas Sb and Zn seemed to be good digesting metals – 75 wt.% Mo recovered. A complete recovery for Mo under argon was never obtained. The samples obtained with Sn showed a 2 mm thick dark grey layer at the salt-metal interface. With Sb and Zn, we had a dark grey mass on one point at the interface. An analysis by SEM of these samples allows us to conclude that all the metallic particles not digested settled through the salt and stayed in the dark grey accumulations. The presence of this dark grey layer and its importance are a sign of how the digestion occurred.

We also carried out some experiments varying the time of contact between the molten salt and the liquid metal. Those results are shown in Figure 5. It seems that time does not allow to eliminate the problems that limit the digestion.

Moriyama, *et al.* [6-8] observed a similar dark grey layer at the interface in their study of actinides/lanthanides separation. They had the same problem of digestion with the inter-metallic compounds formed in the salt between the species reduced and the metal solvent. By studying the influence of temperature, reducing agent concentration and salt composition, they found that the extraction of the wanted elements was not controlled by the reduction but by the mass transfer of the metallic species at the interface. The authors found that more than 15 h were necessary to get the metallic compounds into the metal solvent.

Experiments with larger Mo particles (30 µm) gave good recovery (~80% Mo recovered) for all the digesting metals. This suggests the presence of an oxide film on the metal surface which prevents the Mo particles from being in contact with the metal, especially in the case of Sn. Furthermore, an AUGER spectrometry analysis of the Mo particles showed a superficial oxidation of Mo in MoO$_3$. These two phenomena entail a bad wettability of Mo by the metallic solvent and a good wettability of Mo by the fluorides under argon, and thus a limited digestion. In the case where inter-metallic compounds can be formed between Mo and the metal at the interface, wetting could begin on some points of the particle and then allow the digestion under argon. Sb, Sn and Zn may form inter-metallic compounds with Mo but the kinetics must be different.

Results under hydrogen

In order to see if the absence of the layer of oxide on the liquid metal and around the Mo particles permits to digest these particles, we conducted the same experiments under hydrogen, which is supposed to reduce all these oxides. The results are displayed in Figure 6.

Processing by hydrogen during the digestion seems to solve the problems whatever the digesting metal. It must be noted that, in none of the digesting metals used here, Mo is soluble. Consequently, solubility is not useful for the digestion process. Once the contact is established between the particles and the metal by reduction of the oxide films on each one, the particles must settle into the liquid metal. The formation of inter-metallic compounds may facilitate the wetting, and thus the digestion.

Conclusion

The problem of the extraction of tellurium, noble metals and other easily reducible species (Sn, Fe, Ni,...) from fission products is solved by a volatilisation of tellurium metal and a digestion of the other elements in a metallic phase under hydrogen. For the digestion process, experiments must be performed to clarify the mechanism which lead to a complete extraction of the metallic particles contained in a fluoride salt by a liquid metal. We saw that wettability of the metallic particles by the fluoride salt and by the liquid metal is determinant but all the phenomena have not been discovered. More precisely, agitation and temperature must be important. Kinetics should also be studied.

REFERENCES

[1]   J.A. Jensen, A.M. Platt, G.B. Mellinger, W.J. Blorklund, "Recovery of Noble Metals from Fission Products", *Nucl. Technol.*, 65, 305, 1984.

[2]   K. Naito, T. Matsui, Y. Tanaka, "Recovery of Noble Metals from Insoluble Residue of Spent Fuel", *J. Nucl. Sci. Technol.*, 23, 6, 540-549, June 1986.

[3]   K. Naito, T. Matsui, H. Nakahira, M. Kitagawa, H. Okada, "Recovery and Mutual Separation of Noble Metals from the Simulated Insoluble Residue of Spent Fuel", *J. Nuclear Materials*, NLD, 184, 1, 30-38, 1991.

[4]   H. Moriyama, K. Kinoshita, T. Seshimo, Y. Asaoka, K. Moritani, Y. Ito, "Pyrochemical Recovery of Fission Product Noble Metals", Third International Conference on Nuclear Fuel Reprocessing and Waste Management, RECOD'91, 2, 639-643, 1991, Japan Atomic Ind. Forum, Tokyo, Japan.

[5]   F. Lemort, "Étude de la Séparation Actinides-Lanthanides des Déchets Nucléaires par un Procédé Pyrométallurgique Nouveau", Thèse INPG 1997.

[6]   H. Moriyama, K. Kinoshita, Y. Asaoka, K. Moritani, Y. Ito, "Equilibrium Distributions of Actinides and Fission Products in Pyrochemical Separation Systems, (II)", *J. Nucl. Sci. Technol.*, 27, 10, 937-943, October 1990.

[7]   J. Oishi, H. Moriyama, K. Moritani, S. Maeda, M. Miyasaki, Y. Asaoka, "Behavior of Several Lanthanide and Actinide Elements in a Molten Salt/Liquid Metal Extraction System", *J. Nucl. Mat.*, 154, 163-168, 1988.

[8]   H. Moriyama, M. Miyazaki, Y. Asaoka, K. Moritani, J. Oishi, "Kinetics of Reductive Extraction of Actinide and Lanthanide Elements from Molten Fluoride into Liquid Bismuth", *J. Nucl. Mat.*, 182, 113-117, 1991.

Figure 1. Free enthalpy of various oxides contained in the fission products

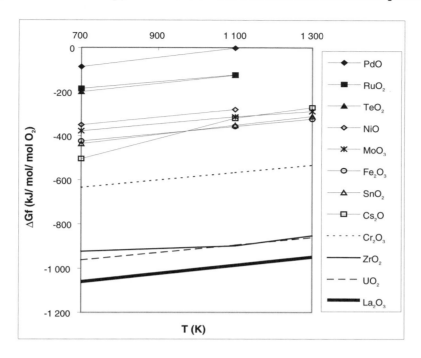

Figure 2. Process of extraction of noble metals and easily reducible species

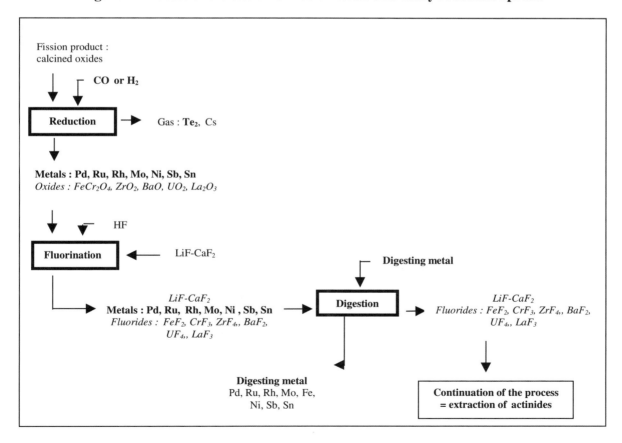

Figure 3. Lost of substances expressed in percentage of initial TeO₂ during the processing by CO

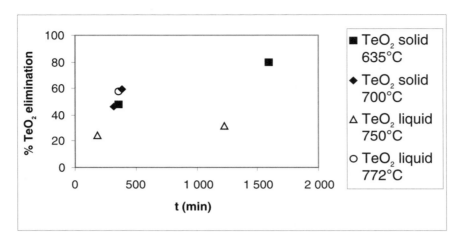

Figure 4. Reactor used in the study of digestion of metallic particles by a liquid metal from a fluoride salt

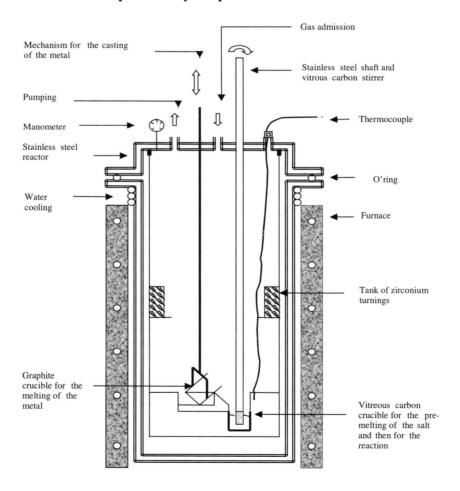

Figure 5. Data of kinetics of the "digestion" of Mo metallic particles

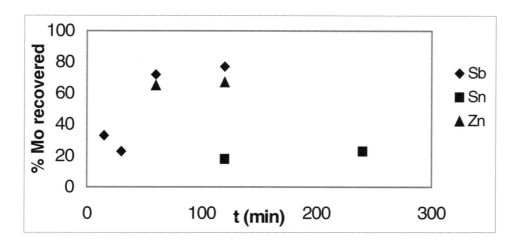

Figure 6. Digestion of Mo particles under argon and under hydrogen

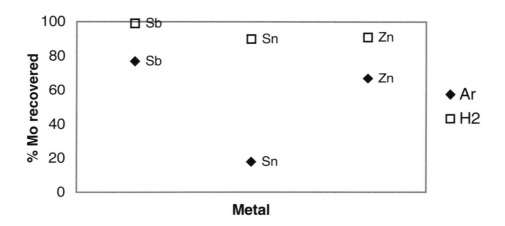

STUDY OF OXIDISING URANIUM FOR PYROCHEMICAL REPROCESSING

Shinichi Kitawaki, Naoki Kameshiro, Mineo Fukushima, Munetaka Myochin
Japan Nuclear Cycle Development Institute
4-33, Muramatsu, Tokai-mura, Naka-gun, Ibaraki, 319-114, Japan
Tel.: 81-029-282-1111 • Fax: 81-029-282-0864

Abstract

To apply the metal electrorefining method developed at Argonne National Laboratory (ANL) to MOX fuel reprocessing, it is necessary to devise an oxidising metal deposit technology. In this paper, a thermodynamics simulation result using MALT2 and a roasting oxidation experiment and solvent cadmium oxidation experiment are discussed. In this experiment, it was confirmed that a reaction in solvent cadmium oxidation is more controllable than other methods. Using a U metal test in solvent cadmium oxidation, it was also established that the oxidation product is UO_2. The oxidising method is confirmed to be able to be applied to the oxidation process in pyrochemical reprocessing.

Introduction

Japan Nuclear Cycle Development Institute (JNC) expects pyrochemical processing to be used as the future reprocessing/fuel fabrication technology because it is a simple process and has the possibility of reducing construction costs as compared with the PUREX/pellet process. JNC is studying the application of the metal electrorefining method developed at ANL to oxide fuel reprocessing.

The process in which oxide fuel is reduced to metal has been developed at ANL and the Central Research Institute of Electric Power Industry (CRIEPI) in Japan. The oxidising process has not been studied because the oxidising reaction goes on by itself with oxygen in air. However, the oxidising process is important to produce fuel material suitable for fuel fabrication. Thus it is very important to discuss the oxidising process for improvement of pyrochemical processing.

Selection of oxidation technology

The oxidation method can be classified into three types (the solid-gas reaction, the solid-liquid reaction, and the gas-liquid reaction) resulting from the reactive system of U metal and oxygen. To apply oxidation technology as a part of pyrochemical reprocessing, its applicability was evaluated and the most effective method was selected.

Solid-gas reaction

The oxidation by solid-gas reaction is a method in which massive U is exposed to gas. The reaction product is changed by oxidant. Since the powdered U metal is usually pyrophoric and burns in air [1], roasting oxidation, which uses massive U, is appropriate for its oxidation.

Solid-liquid reaction

The oxidation by solid-liquid reaction uses water as an oxidant. The oxidation reaction of boiling water and a massive U finally generates UO_2. A similar reaction occurs in water at 25°C in the case that the form of U is a powder. In this reaction, which uses water, hydrogen gas is generated, and can be used as a reducing gas of the O/M adjustment. But the premise of pyrochemical reprocessing (it being a non-water process) was not satisfied. Therefore, this method was removed from selection.

Gas-liquid reaction

The oxidation by gas-liquid reaction is a method in which oxygen is introduced directly into the melting U metal, or U dissolved in low melting metal. Considering the high melting point of metal U, the method of oxidation after dissolution of metal U into low melting metal is appropriate.

The evaluation result of the advantages and faults of each reaction is that two methods, direct oxidation of massive U (roasting oxidation) and oxidation of U dissolved in low melting metal (solvent cadmium oxidation), were selected as candidates for the oxidation process in pyrochemical reprocessing.

Thermodynamics simulation

To examine the oxidation behaviour of the U metal, two oxidation methods were calculated using the MALT2 thermodynamics database [2]. In the thermodynamics calculation, oxidation behaviour was expected from the standard Gibbs free energy of oxide of cadmium and U. Figure 1 shows the temperature dependency of the standard Gibbs free energy of U and cadmium. Solvent cadmium has a tendency of being reduced easily compared with U. Therefore, at oxidising U in solvent cadmium, UO_2 was the limiting oxidation state even as cadmium underwent some oxidation. Moreover, the standard Gibbs free energy of U_3O_8 becomes smaller than that of UO_3, more than 594°C. U_3O_8 is the most stable oxide at this temperature. Each oxidation method was simulated based on the calculated standard Gibbs free energy. In this thermodynamics simulation, oxidation behaviour was calculated with feed oxygen (1-3 moles) and reaction temperature (25-594°C) as variables.

Figure 2 shows the simulation result of the roasting oxidation method. The oxidation reaction progresses and U oxidation order is UO_2, U_4O_9, U_3O_8, and UO_3 as the amount of feed oxygen increases. Figure 3 shows the simulation result of the solvent cadmium oxidation method. The oxidation order was as well as the roasting oxidation. However, when cadmium exists, it was confirmed that the U metal did not progress to a further oxidation until the end of generation of cadmium oxide. In the result of thermodynamics simulation, the last oxidation product was U_3O_8 at more than 594°C, UO_3 at less than 594°C. Moreover, product material by the oxidation reaction became in order of U, UO_2, UO_{2+x}, U_3O_8, and UO_3. The chemical composition has changed depending on the amount of oxygen supplied.

Roasting oxidation experiment

To determine the experiment apparatus material, the corrosion resistance was calculated using the thermodynamics database. In this calculation, Ta, Ti and Al_2O_3 and ZrO_2 were selected as apparatus material. As a result, a ZrO_2 crucible was selected based on its chemical stability. Figure 4 shows the outline of the experiment apparatus. The crucible made of ZrO_2 was put in the protection container made of stainless steel. Columnar metal U (ϕ 6 × 50) of a test piece was put on the glass stage. As for the glass stage, there are some holes so that oxygen gas may flow. In addition, to make oxygen gas react with U efficiently, the blowing tube was set up in the crucible bottom.

Table 1 shows the examination conditions. The reaction temperature was 300°C, 500°C; the flow rate of reactant gas was 100 ml/min, 1 000 ml/min; reaction time was 5-24 hours, and the oxygen concentration of reactant gas was 10% and 20%. To evaluate converted U oxide after examination, appearance observation, weight measurement, and X-ray diffraction of the sample were performed. The apparent oxidation rate was evaluated from the weight change and the thickness of oxide film using the following equation:

> Apparent oxidation rate = Thickness of oxide film/time
> Thickness of oxide film = amount of oxide/(specific gravity × sample surface area)

Table 2 shows the examination result. To discuss the influence on apparent oxidation rate by inflow of reactant gas, RUN-1 was compared with RUN-2. The oxidation rate of RUN-1 (100 ml/min) was 46.08 μm/h, and that of RUN-2 (1 000 ml/min) was 16.03 μm/h. The reason for the apparent decrease in oxidation rate during RUN-2 is guessed to be that the surface of metal U is cooled with feed gas which is non-preheated.

To discuss the influence of reaction temperature on the apparent oxidation rate, RUN-1 (five hours at 300°C) and RUN-3 (24 hours at 300°C) were compared with RUN-4 (five hours at 500°C) and RUN-5 (24 hours at 500°C) respectively. The result of RUN-4, 212.19μm/h, was five times as large as the result of RUN-1, 46.09 μm/h. Moreover, the apparent oxidation rate of RUN-5 was 58.13 μm/h, and that of RUN-3 was 9.2 μm/h. The apparent oxidation rate of RUN-5 was six times as large as that of RUN-3. Therefore, in this examination system, there is a tendency of the apparent oxidation rate to increase at the same reaction time when the reaction temperature rises.

In addition, RUN-1 and RUN-3 had almost the same oxide ratio (= oxidised U/initial metal $U \times 100$), being 17.8%, and 16.2% respectively. At 300°C, it was confirmed that oxidation does not progress after five hours. In the examination result to confirm the oxidation behaviour by reaction time, the apparent oxidation rate was 201.2 μm/h at five hours (RUN-8), 139.2 μm/h at seven hours (RUN-9), 198.2 μm/h at ten hours (RUN-10) and the oxide ratio was 71%, 100% and 100% respectively. Therefore, it was confirmed that the U metal was completely oxidised in seven hours at 500°C.

The oxidation product properties did not depend on reaction time, and there was a particle size distribution of 50-200 μm. The chemical form was a compound of UO_2, UO_3 and U_3O_8, and O/M was 2.62-2.68.

Solvent cadmium oxidation experiment

The corrosion resistance of the apparatus material for solvent cadmium experiment was discussed as well as the roasting oxidation experiment to make the experiment device, and Al_2O_3 was selected as a crucible material. Figure 5 shows the outline of the experimental apparatus. The crucible made of Al_2O_3 was put in the protection container made of quartz, and to make the oxygen gas react with U efficiently, the blowing tube was set up in the crucible bottom.

Table 3 shows the examination conditions. The U metal in cadmium was 4.76 wt.%, which was twice as large as solubility. In this experiment, the $Ar-1\%O_2$ gas was blown in for 1-4 hours. The flow rate was 500 ml/min and 1 000 ml/min, and temperature was 600°C. After the experiment, converted U oxide was evaluated by appearance observation, X-ray diffraction and ICP-AES.

In the examination, the influence of oxidation time (equivalent of oxygen) was confirmed. Table 4 shows the examination result. In RUN-1 the blank sample was measured and the correction value was decided to evaluate the examination result. In this examination, metal U oxidised completely in four hours (amount of the feed oxygen is two equivalents). There was a proportional relation between the oxidation time and the oxide ratio, though U was slowly oxidised for one hour of oxidation beginning. According to X-ray diffraction measurement, the products were UO_2 and CdO. In RUN-4, the most of U was oxidised in solvent cadmium, although a part of U had changed into CdU_2O_7.

Discussion

Comparison with thermodynamic calculation result

In the roasting oxidation experiment, the temperature influences oxidation rate. Figure 6 shows the relation between reaction temperature and apparent oxidation rate. The differential of apparent oxidation rate is large between 200°C and 400°C, though it is small at less than 200°C and larger than

400°C. It is thought that the rate-determining step of oxidation reaction is different in three temperature ranges. The rate-determining step at less than 200°C may be diffusion of oxygen in uranium metal and that between 200-400°C may be oxidation to U_3O_8. At more than 400°C it may be oxidation to UO_3. Therefore, using ingot U, it is difficult to generate only UO_3 by feed gas control.

In the solvent cadmium oxidation experiment, the oxide ratio is influenced by the amount of feed oxygen. Figure7 shows the relation between oxidation time and oxide ratio. U was oxidised by 40% in two hours (about one equivalent), and it was detected that a part of cadmium was oxidised. The cause is guessed to be an insufficient stirring. The entire reaction product was UO_2, as is shown by the thermodynamics calculation, and the reaction of UO_{2+x} did not take place. Though U more than solubility was added, the whole product was UO_2. Therefore, even if the U does not dissolve completely in cadmium, a further oxidation of UO_2 could be restricted by the existence of cadmium.

Feasibility of oxidation process

Comparing the examination results with calculation, it was shown that the oxidation form management is difficult only by controlling the temperature and the amount of feed gas in roasting oxidation. On the other hand, it was shown that there was a possibility that the oxidation form could be controlled with the amount of feed gas in solvent cadmium oxidation. Figure 8 shows the process flow when two oxidation technologies are adopted to a part of the pyrochemical reprocessing. In these cases, the fuel form might be pellet. The feature of each process is as follows. If the roasting oxidation is adopted, the starting material (U and/or Pu) has to be ingot processed by cathode processor, because of the possibility of metallic fire. Since the oxidation product has large O/M, a reduction process such as hydrogen reduction is needed. When the solvent cadmium oxidation method is adopted, cathode deposition obtained by electrorefining directly becomes the starting material of oxidation. Its oxidation form is only UO_2, rendering post-processing unnecessary. However, the product material becomes a fine powder which includes a small amount of CdO, so handling will be difficult.

Conclusion

It was confirmed that two methods could oxidise U metal completely. Moreover, there was no problem regarding safety. As for the roasting oxidation method, it was confirmed that it was difficult to generate the oxide with a single chemical form. On the other hand, the solvent cadmium oxidation method generates UO_2 and a little CdO. The feasibility of the oxidation method for pyroprocess was confirmed in this result. However, only U was targeted in this discussion. The oxidation behaviour of U/Pu compounds should also be studied.

REFERENCES

[1]   J.J. Katz, E. Rabinowitch, "The Chemistry of Uranium", 165-167, Dover Publications.

[2]   Materials-Oriented Little Thermodynamic Database (MALT2).

Figure 1. Temperature dependency of Gibbs free energy

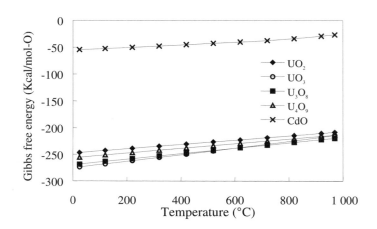

Figure 2. Thermodynamic simulation result of roasting oxidation

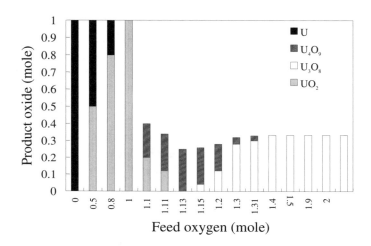

Figure 3. Thermodynamic simulation result of solvent cadmium oxidation

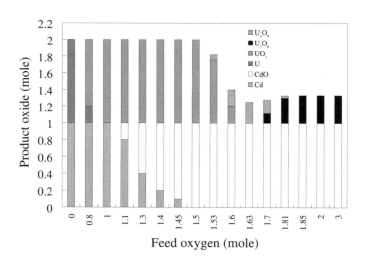

Figure 4. Outline of roasting oxidation experiment apparatus

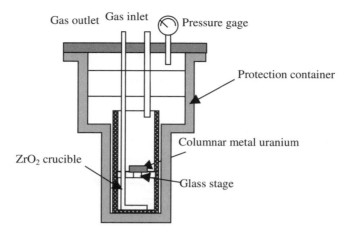

Figure 5. Outline of solvent cadmium oxidation experiment apparatus

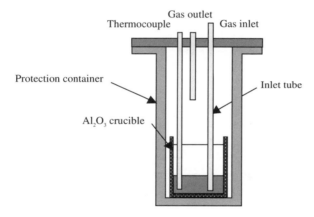

Figure 6. Relation between temperature and oxidation rate

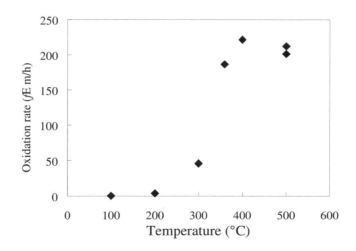

Figure 7. Relation between oxide ratio and time

Oxidation rate U
Oxidation rate Cd
Oxide ratio (%)
Time (hour)
100
90
80
70
60
50
40
30
20
10
0
0
0.5
1
1.5
2
2.5
3
3.5
4

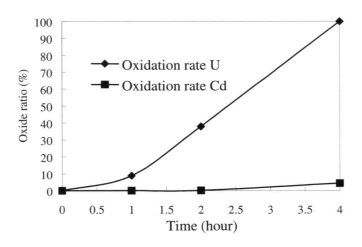

Figure 8. Pyrochemical reprocessing flow contains oxidation process

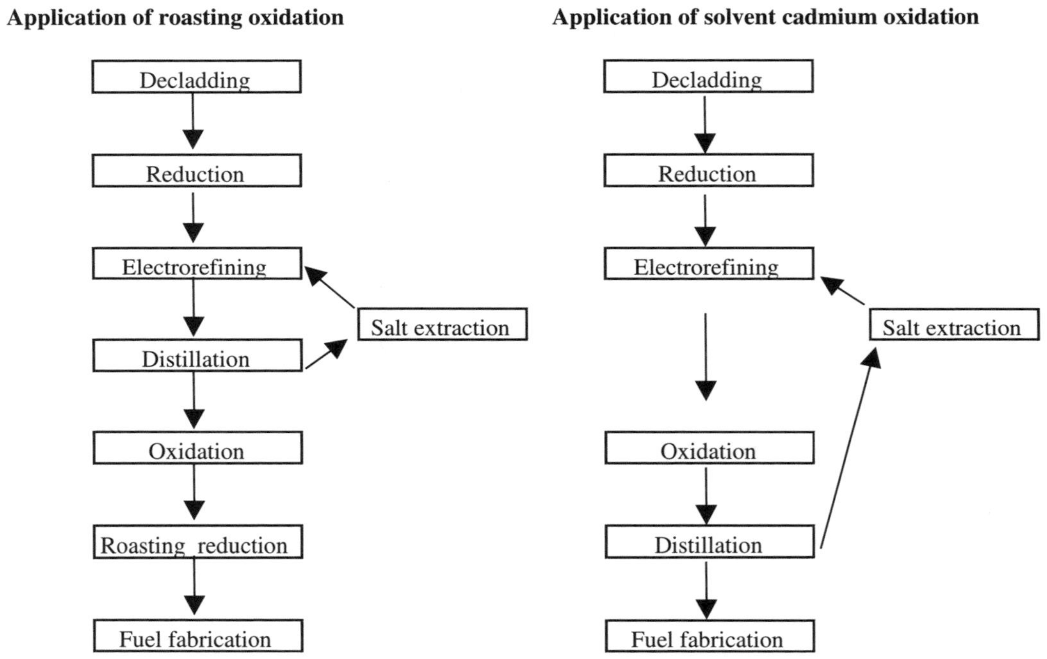

Application of roasting oxidation
Decladding
Reduction
Electrorefining
Distillation
Oxidation
Roasting reduction
Fuel fabrication
Salt extraction
Application of solvent cadmium oxidation
Decladding
Reduction
Electrorefining
Oxidation
Distillation
Fuel fabrication
Salt extraction

Table 1. Roasting oxidation examination

RUN No.	Temperature (°C)	Flow rate (ml/min)	Reaction time (hour)	Concentration of O_2 (%)
1	300	100	5	20
2	300	1 000	5	20
3	300	100	24	20
4	500	100	5	20
5	500	100	24	20
6	500	100	5	10
7	300	100	5	10
8	500	100	5	20
9	500	100	7	20
10	500	100	10	20

Table 2. Roasting oxidation examination

RUN No.	Initial weight of metal U (g)	Weight of converted U (g)	Apparent oxidation rate (μm/h)	Oxide ratio (%)	Oxidation product
1	24.415	4.35	46.08	17.8	UO_2, UO_3, U_3O_8
2	26.195	1.593	16.03	6.1	UO_2, UO_3, U_3O_8
3	25.713	4.168	9.2	16.2	UO_2, UO_3, U_3O_8
4	29.285	20.032	212.2	68.4	UO_2, UO_3, U_3O_8
5	26.312	26.312	58.13	100	UO_2, UO_3, U_3O_8
6	26.017	18.201	192.8	70	UO_2, UO_3, U_3O_8
7	26.628	2.238	23.71	8.4	UO_2, UO_3, U_3O_8
8	53.279	37.993	201.2	71	UO_2, UO_3, U_3O_8
9	52.391	52.391	198.2	100	UO_2, UO_3, U_3O_8
10	52.581	52.581	139.2	100	UO_2, UO_3, U_3O_8

Table 3. Solvent cadmium oxidation examination

RUN No.	Temperature (°C)	Flow rate (ml/min)	Reaction time (hour)	Concentration of O_2 (%)
1	600	0	0	1
2	600	500	1	1
3	600	500	2	1
4	600	500	4	1

Table 4. Solvent cadmium oxidation result

RUN No.	Uranium oxide ratio(%)			Cadmium oxide ratio (%)		
	Measurement value	Corrected value	Oxidation form	Measurement value	Corrected value	Oxidation form
1	9	0	UO_2	0.1	0	CdO
2	18	9	UO_2	0.1	0.1	CdO
3	47	38	UO_2	0.3	0.2	CdO
4	100	100	UO_2	4.6	4.5	CdO

PARTITIONING OF AMERICIUM METAL FROM RARE EARTH FISSION PRODUCTS BY ELECTROREFINING

Carole Pernel, Michel Ougier, Jean-Paul Glatz, Lothar Koch
Institute for Transuranium Elements
Postfach 2340, D-76125 Karlsruhe, Germany

Tadafumi Koyama
Central Research Institute of Electric Power Industry
2-11-1 Iwadokita, Komae-shi, Tokyo 201-0004, Japan

Abstract

The present study is focused on the separation of minor actinides from lanthanides by electrorefining in a LiCl-KCl melt. Special attention is given to the recovery of americium, which seems to be the most difficult to separate from lanthanides for two main reasons. Firstly because, of all actinides, americium chlorides have the closest free standard energy of formation to lanthanides. Secondly because americium is a multiple valency element in chloride media; in LiCl-KCl it can be present in the oxidation states +2 and/or +3. In fact, the americium valency in molten chloride and fluoride salts is still unclear and contradictory results are found in literature. Therefore, an experimental set-up has been developed at ITU to extensively study these phenomena in order to optimise the americium electrorefining.

Scope

There is presently a surplus of low cost U ores, which could fuel the presently deployed light water reactor (LWR) for the next 50 years. Moreover, enriched U released from military stockpiles is gradually becoming available. Despite the low use of U for electricity generation in LWR of about 1% compared to the fast breeder reactor (FBR) – which would improve the use of U by a factor 80 – the current high cost of Pu separation makes this reactor type non-competitive with the cheap LWR fuel. If the greenhouse effect restricts the future use of fossil fuels and if no other substitutes for energy generation exist, then necessarily the use of nuclear energy and the demand for U would increase, and consequently fast reactors would be introduced sooner. In the event of such a scenario, a partitioning and transmutation (P&T) concept could be realised which could reduce the long-term radiotoxicity of the nuclear waste. Although suitable aqueous processes have been developed to treat spent nuclear fuel by separating out, not only Pu-U (PUREX process) [1], but also all long-lived radiotoxic nuclides (DIAMEX, TRUEX, SANEX) [2], pyrochemical reprocessing is advantageous. Due to its compactness, such a process could be attached to each power station, thus reducing the need for spent fuel transportation to a central large scale reprocessing plant. Also, a pyrochemical process would allow a speedy reprocessing, which would eliminate the out of pile fuel inventory and reduce the [241]Am built up due to [241]Pu decay.

There have been two pyrochemical processes developed for the separation of Pu-U from spent fuel by electrolysis in molten chlorides, firstly as metals [3-5], and secondly as oxides [6,7]. In order to extend the application to a P&T scheme, they have to be supplemented with the separation of minor actinides (MA). Neptunium, with chemical properties similar to U and Pu, is not difficult to separate. However, the transplutonium elements exhibit standard free energies or electropotentials, which are close to the rare earth elements (RE) (cf. Figure 2), and the MA/RE separation seems difficult.

This inherent difficulty in oxide electrolysis lead Bychkov, *et al.* to redesign the process by including a fractionated precipitation of transplutonium oxides (Dovita process) in order to separate these from the rare earth elements [8]. The situation in the electrorefining of minor actinide metals is similar. Studies carried out by CRIEPI at Missouri University show that the complete separation of Am from rare earth elements seems to be an insurmountable obstacle. Their solution is an initial electrorefining of U and Pu followed by a multi-stage reductive extraction of the minor actinides, particularly americium, from rare earths [9,10] (CRIEPI process). In neither case are the fractional precipitation of the oxides and the multi-stage extraction of americium metal the ideal solution compared to an electrochemical separation of americium together with the other actinides. Therefore, the objective of our study is to achieve an americium separation yield of 99.9% only by electrolysis process.

Discussion of the electrorefining process

In a P&T scheme, a separation yield of 99.9% for each minor actinide has to be achieved if the radiotoxicity of the remaining nuclear waste is to be reduced by a factor of about 500 times. So far, it seems to be impossible to reach this goal through oxide electrolysis alone. The relevant basic data for americium dioxide are not known. In case of the electrolysis of minor actinide metals, experimental studies on a milligram scale are promising except for americium [10]. Possible explanations are firstly the small difference between the free standard energy of lanthanide and actinide elements and secondly the redox reaction of the deposited americium with americium trichloride in the melt according to the equation [10,11]:

$$Am + 2AmCl_3 \Leftrightarrow 3AmCl_2 \tag{1}$$

The published relevant thermodynamic data, which support this reaction, are still lacking. In the following paragraph the known thermodynamic data will be reviewed in order to determine the conditions that stabilise americium trichloride in the electrolyte.

Americium recovery is the most difficult in the process of actinides/lanthanides separation, because americium halides have the closest standard free energy of formation to lanthanides of all actinides. This has been demonstrated by electrochemical experiments with chlorides [9,11-13] and predicted thermodynamically for fluorides [14,15] (see Figure 2). It is important to note that ΔG_f^0 of americium chloride is closer to ΔG_f^0 of plutonium chloride than to ΔG_f^0's of lanthanide chlorides. For the fluorides, it is the opposite. Consequently, the free energy values of lanthanides and americium in chloride media show a greater separation than in fluoride media. Hence, the americium/lanthanides separation by electrorefining in salt melt should be easier in chlorides than in fluorides, not to mention the lower melting point of the proposed chloride eutectic mixture [16]. Figure 2(a) presents americium as dichloride whereas Figure 2(b) presents it as a trifluoride. In fact, the americium valency in molten chloride and fluoride salts is in question, and contradicting results are found in the literature, which are summarised in Table 1.

The multi-valency of americium prevents a sufficiently high recovery yield by electrorefining. If the above Eq. 1 is valid, a solution to this problem would be to shift the equilibrium in favour of americium trichloride, which has a predicted potential more separated from the potential of lanthanide trichloride than americium dichloride. Considering the mass law [29]:

$$K = \frac{a^3_{Am(II)}}{a_{Am(0)} a^2_{Am(III)}} \tag{2}$$

where a_x is the activity of the compound X which is equal to the product of the activity coefficient γ and the concentration c of X:

$$a_x = \gamma_x \cdot c_x \tag{3}$$

Americium trichloride is favoured if the activity of americium metal is decreased. Hence, according to Eq. 3, one has to reduce the activity coefficient or the concentration. The use of a liquid metal (Cd or Bi) cathode reduces the activity coefficient of americium metal drastically. Moreover, there are no losses of the deposited metal due to its poor adherence to a solid cathode. Americium metal when formed at the surface of the liquid cathode is transported directly into the bulk of the cadmium, limiting the reaction between americium metal and americium trichloride (Eq. 1). In order to eliminate this reaction at all, the liquid cathode should be continuously withdrawn. Consequently, as in the process for residual U and Pu metal recovery (Figure 1), the most suitable cathode for americium recovery by electrorefining will be a liquid metal cathode.

One can suppose that at any time the molten salt phase is in thermodynamic equilibrium with the liquid metal composing the cathode. If this assumption is valid, then the main results of the liquid extraction experiments in molten media would provide very useful information for determining the optimal conditions for americium electrorefining. In order to compare the distribution coefficients D_x for individual element X, they have been calculated (Figure 3) using published separation factors $SF_{Nd}(X)$ of a reductive extraction, where, from a molten LiCl-KCl eutectic mixture at 773 K, the actinides are extracted into Cd [12,17,18] or Bi [17,18] following a reduction with Li. It is seen that there are larger separation factors ($SF_{Nd}(X) = D_x/D_{Nd}$) for Bi compared to Cd extraction. This is further supported by activity coefficient measurements (Table 2), which show that americium has a lower activity coefficient in liquid Bi than in liquid Cd. This, in turn, exhibits that americium metal is more stable and hence more easily separated from the salt melt by liquid Bi. To further improve the electrorefining process of actinide metals, it would be worthwhile to replace the Cd cathode by Bi.

The concept to partition americium metal from rare earth fission products

Theoretically, according to the above considerations, an efficient way to improve the americium recovery by electrorefining would be a continuous withdrawal of the liquid cathode by airlift or pumping, which would be treated by an evaporation-condensation cadmium cycle. The set-up of such a complex process is under development. In a first step, the approval will be tested by an alternative exchange of the liquid cathode in the course of the electrolysis before the more complex process is studied. Before experimenting with a continuous renewal of the liquid metal cathode, it is necessary to test a liquid metal cathode in which the americium activity coefficient is lower than in Cd, i.e. the liquid bismuth cathode.

These experiments require extreme conditions to handle even milligram quantities. The high radiotoxicity and specific activity of americium necessitate the use of a α-tight hot cell or glove box with biological shielding to protect personnel from β-γ radiation and radionuclide incorporation. Furthermore, due to the hygroscopic nature of LiCl-KCl actinide oxide ions and oxychlorides can be formed, if a certain amount of water is absorbed. In addition, traces of oxygen can interfere with the process by oxidising certain elements and by forming emulsions at the molten salt/molten metal interface [34], which hampers the exchange reaction between the two phases. Therefore, the experiments must be performed in an inert gas atmosphere.

Two specific installations for experimental pyrometallurgy have been designed and constructed at the Institute for Transuranium Elements (ITU). The larger one, which enables to carry out experiments at a few tenths of grams scale, is a stainless steel box ("caisson"), for the pyrometallurgical processing of metal fuel and HLLW in the frame of a ITU-CRIEPI contract and is presented in another paper of this workshop [35]. The smaller installation consists of a double glove box dedicated to the above mentioned americium study at the 100 mg scale. Figure 4 shows the complete system comprising a double glove box and an argon purification unit. The pyrometallurgical experiments will be carried out inside a heating well of 100 mm diameter fitted into the bottom of the inner glove box, which is under argon atmosphere. An upper water-cooled instrumentation flange enables an accurate positioning of the electrodes, the thermocouple and the stirrer; it also supports the deflector and the electrorefiner (Figure 5). The whole assembly can be moved up and down by means of a lifting system. The required oxygen and water concentrations in the inner box, less than ten ppm each, are maintained by metal catalyst and molecular sieve traps in the argon purification system. The outer box, which is under nitrogen atmosphere, is dedicated to the sample preparation for analysis.

Basic electrochemical measurements on lanthanides will be carried out. To improve the handling and to assess the reproducibility and accuracy of the method, the first experiments will be carried out with lanthanum and neodymium, the latter having a similar behaviour to americium. The electrorefining of americium will be performed under different conditions (with various ratios of lanthanide/americium concentrations, with or without metallic pool (Cd, Bi) and with various cathodes (Cd, Bi)) in order to define the limiting conditions and to optimise recovery yields.

REFERENCES

[1]   J. Malvyn, Mckibben, *Radiochim. Acta*, 36, 3 (1984).

[2]   "Actinide Separation Chemistry in Nuclear Waste Streams and Materials", Nuclear Energy Agency, December (1997).

[3]   J.P. Ackerman, *Ind. Eng. Chem. Res.*, 30 (1), 142 (1991).

[4]   T. Koyama, M. Iizuka, Y. Shoji, R. Fujita, H. Tanaka, T. Kobayashi, M. Tokiwai, *J. of Nuc. Sci. and Tech.*, 34, N. 4, 84 (April 1997).

[5]   M. Iizuka, T. Koyama, N. Kondo, R. Fujita, H. Tanaka, *J. of Nuc. Mat.*, 247 183 (1997).

[6]   V.B. Ivanov, O.V. Skiba, A.A. Mayershin, A.V. Bychkov, L.S. Demidova, P.T. Porodnov, Proc. of GLOBAL'97, 2, 906.

[7]   A.V. Bychkov, S.K. Vavilov, P.T. Porodnov, A.K. Pravdin, G.P. Popkov, K. Suzuki, Y. Shoji, T. Kobayashi, Proc. of Global'97, 2, 912.

[8]   A.P. Kirilloovich, A.V. Bychkov, O.V.S Skiba, L.G. Babikov, Yu.G. Lavrinovich, A.N. Loukinykh, Proc. of GLOBAL'97, 2, 900.

[9]   Y. Sakamura, T. Hijikata, K. Kinoshita, T. Inoue, T.S. Storvick, C.L. Krueger, J.J. Roy, D.L Grimmett, S.P. Fusselman, R.L Gay, *J. of Alloys and Compounds*, 271-273, 592 (1998).

[10] Y. Sakamura, *et al.*, Proc. of GLOBAL'95, 2, 1185.

[11] K. Uozumi, Y. Sakamura, K. Kensuke, S.P. Fusselman, C.L. Kueger, CRIEPI-Report T98011 (1999).

[12] T. Koyama, T.R. Johnson and D.F. Fischer, *Journal of Alloys and Compounds*, 189, 37 (1992).

[13] M. Kurata, Y. Sakamura, T. Matsui, *J. Alloys and Compounds*, 234, 83 (1996).

[14] I. Barin, VCH Verlags Gesellshaft, Weinheim (1989).

[15] A. Glassner, USAEC Report, ANL-5750, (1958).

[16] G.J. Janz, "Molten Salts Handbook", Academic Press Inc. (1967).

[17] M. Kurata, Y. Sakamura, T. Hijikata, K. Kinoshita, *J. Nucl. Mater.*, 227, 110 (1995).

[18] K. Kinoshita, T. Inoue, S.P. Fusselman, D.L. Grimmet, J.J. Roy, C.L. Krueger, C.R. Nabelek, T.S. Storvick, *J. Nucl. Sci. Technol.*, 36 (2), 189 (1999).

[19]   L. Martinot, J.C. Spirlet, G. Duyckaerts, W. Muller, *Bull. Soc. Chim. Belges*, 76, 211 (1967).

[20]   J.L. Mullins, A.J. Beaumont and J.A. Leary, *J. Inorg. Nucl. Chem.*, 30, 247 (1968).

[21]   J.C. Mailen, L.M. Ferris, "Distribution of Transuranium Elements Between Molten Lithium Chloride and Lithium Bismuth Solutions: Evidence for Californium (II)", *Inorg. Nucl. Chem. Letters*, 7, 431 (1971).

[22]   L.M. Ferris, F.J. Smith, J.C. Mailen, M. Bell, *Journal of Nuclear Chemistry*, 34, 2921 (1972).

[23]   L. Martinot, J.C. Spirlet, G. Duyckaerts, W. Muller, *Analytical Letters*, 6 (4), 321 (1973).

[24]   J.A. Leary and L.J. Mullins, *J. Chem. Thermodynamics*, 6, 103 (1974).

[25]   V.P. Kolesnikov, G.N. Kazantsev, O.V. Skiba, *Radiokhimiya*, 19 (4), 545 (1977).

[26]   L. Martinot, "Molten Salt Chemistry of Actinides", Handbook on the Physics and Chemistry of the Actinides, A.J. Freeman and C. Keller, eds., Elsevier Science Publishers B.V. (1991).

[27]   R.I. Gay, L.F. Grantham, S.P. Fusselman, AIP Conference on Accelerator-Driven Transmutation Technologies and Applications, Las Vegas, Nevada, 25-29 July 1994.

[28]   Y. Sakamura, H. Miyashiro, T.S. Storvick, L.F. Grantham, CRIEPI-Report, T94027 (1995).

[29]   Y. Sakamura, T. Inoue, T.S. Storvick, L.F. Grantham, "Development of the Pyropartitioning Process: Separation of Transuranium Elements from Rare Earths Elements in Molten Chlorides Solution: Electrorefining Experiments and Estimations by Using the Thermodynamic Properties", Proceedings of GLOBAL'95.

[30]   K. Kinoshita, T. Inoue, S.P. Fusselman, R.L. Gay, C.L. Krueger, T.S. Storvick, N. Takahashi, Proceedings of GLOBAL'97, 2, 820.

[31]   L.M. Ferris, J.C. Mailen, J.J. Lawrance, F.J. Smith and E.D. Nogueira, *Journal of Nuclear Chemistry*, 32, 2019 (1970).

[32]   J.C. Mailen, L.M. Ferris, *Inorg. Nucl. Chem. Letters*, 7, 431 (1971).

[33]   L.M. Ferris, J.C. Mailen, F.J. Smith, *Journal of Nuclear Chemistry*, 33, 1325 (1971).

[34]   F. Lemort, "Étude de la Séparation Actinides-Lanthanides des Déchets Nucléaires par un Procédé Pyrochimique Nouveau", Thèse INPG (January 1997).

[35]   T. Koyama, K. Kinoshita, T. Inoue, M. Ougier, J.-P. Glatz, L. Koch, "Small Scale Demonstration of Pyrometallurgical Processing for Metal Fuel and HLLW", OECD/NEA Workshop on Pyrochemical Separation (14-15 March 2000).

Table 1. Reported americium valency in different chlorides and fluorides systems

Ref.	Salt phase	Metal phase	Temp. °C	Method used to determine Am valency	Am valency
[19] 1967	LiCl-KCl	No	400-650	Measurement of the EMF of Am/Am(III),LiCl-KCl/Ag(I), LiCl-KCl/Ag	3
[20] 1965	NaCl-KCl	Pu	698-775	Measurements of the equilibrium distribution: Plot of logs D_M vs. $LogD_{ref.}$	2
[21] 1971	LiCl	Bi	640	Measurements of the equilibrium distribution: Plot of logs D_M vs. $LogD_{ref.}$	3
[22] 1972	LiCl	Bi	640-700	Measurements of the equilibrium distribution: Plot of logs D_M vs. $LogD_{ref.}$	2-3
[23] 1973	LiCl-KCl	No	400-650	Chronopotentiometry	3
[24] 1974	NaCl-KCl	No	698-775	Measurements of the equilibrium distribution: Plot of logs D_M vs. $LogD_{ref.}$	2
[25] 1977	NaCl-KCl	No	700	Chronopotentiometry	3
[26] 1991	LiCl-KCl	No	400-650	Review article: Several methods are referred	3
[12] 1992	LiCl-KCl	Cd	500	Measurements of the equilibrium distribution: Plot of logs D_M vs. $LogD_{ref.}$	3
[27] 1994	LiCl-KCl	No	500	Cyclic voltammetry	2-3
[28] 1995	LiCl-KCl	No	500	Measurement of the EMF	2
[29] 1995	LiCl-KCl	No	500	Measurement of the EMF of Am/Am(II), LiCl-KCl/Ag(I), LiCl-KCl/Ag, cyclic voltammetry	2
[30] 1997	LiCl-KCl	Bi	500	Measurements of the equilibrium distribution: Plot of logs D_M vs. $LogD_{ref.}$	3
[9] 1998	LiCl-KCl	No	500	Cyclic voltammetry Measurement of the EMF of Am/Am ion, LiCl-KCl/Ag(I), LiCl-KCl/Ag Electrorefining	2-3
[31] 1970	LiF/BeF$_2$	Bi	500-700	Measurements of the equilibrium distribution: Plot of logs D_M vs. $LogD_{ref.}$	3
[32] 1971	Fluoride	Bi	640	Measurements of the equilibrium distribution: Plot of logs D_M vs. $LogD_{ref.}$	2-3
[33] 1971	LiF-BeF$_2$(ThF$_4$)	Bi	600	Measurements of the equilibrium distribution: Plot of logs D_M vs. $LogD_{ref.}$	2-3
[22] 1972	LiF-BeF$_2$-ThF$_4$	Bi	640-700	Measurements of the equilibrium distribution: Plot of logs D_M vs. $LogD_{ref.}$	2-3
[34] 1997	LiF-CaF$_2$	Zn/Mg	720	Measurements of the equilibrium distribution: Plot of logs D_M vs. $LogD_{ref.}$	3

Table 2. Activity coefficients of actinides and lanthanides in Cd and Bi

Element	Activity coefficient in Cd			Activity coefficient in Bi			Ref.
	673 K	723 K	773 K	673 K	723 K	773 K	
U	6.31E+01	7.94E+01	1.00E+02		7.90E-06		[9]
Np	1.60E-03	4.00E-03	7.90E-03				[9]
Pu	1.10E-05	4.20E-05	1.40E-04				[9]
Am	2.90E-05	1.30E-04			3.90E-10		[9,11]
Cm			3.00E-05				[12]
Nd	3.20E-10	2.00E-09	1.00E-08				[13]
La	6.30E-11	5.00E-10	4.00E-09	4.00E-15	4.00E-14		[13]
Ce	2.00E-10	1.60E-09	1.30E-08	1.30E-14	1.30E-13		[13]
Pr	3.20E-10	2.50E-09	2.00E-08	1.60E-14	1.60E-13	1.00E-12	[13]
Gd	3.20E-08	1.60E-07	6.30E-07	1.60E-12	1.00E-11	4.00E-11	[13]
Y	6.30E-08	2.50E-07	7.90E+07	7.90E-06	2.00E-05		[13]

Figure 1. Schematic flow of electrorefining process for metal

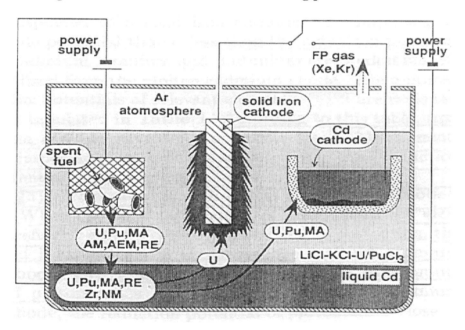

Figure 2. Standard free energy of formation of chlorides and fluorides vs. T

Chloride has been electrochemically determined. Reported liquid fluoride ΔG^0 were extrapolated to lower temperature by the authors to obtain values for super-cooled liquid status.

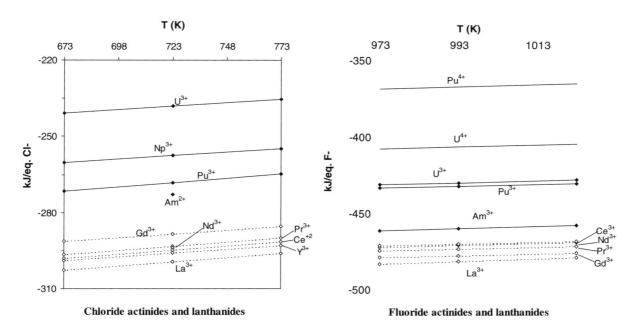

Chloride actinides and lanthanides

Fluoride actinides and lanthanides

Figure 3. Distribution coefficients of actinides and lanthanides in LiCl-KCl/Cd and in LiCl-KCL/Bi where

$$D_X = \frac{\text{molar concentration of } X \text{ in salt phase}}{\text{molar concentration of } X \text{ in metal phase}}, \ D_{Nd} \text{ is an arbitrary reference}$$

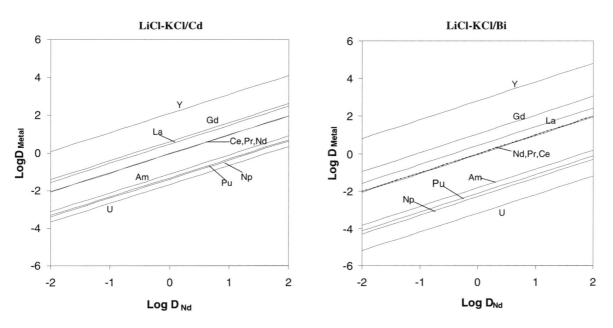

Figure 4. Double glove box with purification unit

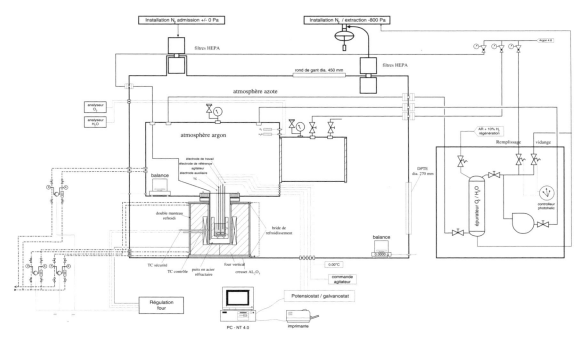

Figure 5. Design of pyrochemical small-scale apparatus

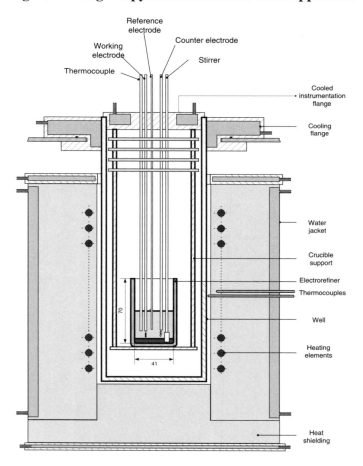

AMERICIUM EXTRACTION VIA IN SITU CHLORINATION OF PLUTONIUM METAL

Wayne A. Punjak, Alfonso J. Vargas, Jr., Gregory D. Bird, Eduardo Garcia
University of California – United States Department of Energy

Abstract

The synthesis of plutonium trichloride via the direct chlorination of molten plutonium has proven to be an effective method for the removal of americium from plutonium metal. Chemical analyses of reactants and products have shown that, at equilibrium, americium preferentially resides in the chloride salt phase. Periodic sampling during chlorination suggests that the approach to equilibrium is very rapid. Based on this assumption, a simplified mathematical model has been developed to predict americium extraction as a function of time, given a specified set of initial conditions and chlorination rate. In this presentation experimental data is compared to model calculations.

PYRO-OXIDATION OF PLUTONIUM SPENT SALTS WITH SODIUM CARBONATE

G. Bourgès, A. Godot, C. Valot, D. Devillard
CEA – Centre d'Études de Valduc
21120 Is-sur-Tille, France

Abstract

The purification of plutonium generates spent salts, which are temporarily stored in a nuclear building. A development programme for pyrochemical treatment is in progress to stabilise and concentrate these salts in order to reduce the quantities for long-term disposal. The treatment, inspired by work previously done by LANL, consists of a pyro-oxidation of the salt with sodium carbonate to convert the actinides into oxides, then of a vacuum distillation to separate the oxides from the volatile salt matrix.

Pyro-oxidation of NaCl/KCl base spent salts first produces a "black salt" which contains more than 97% of the initial actinides. XRD analyses indicate PuO_2 as major plutonium species and sodium plutonates or plutonium sub-oxides PuO_{2-x} can also be identified. Next appears a "white salt" containing less than 500 ppm of plutonium, which meets the operational criterion for LLW discard. For these salts, the pyro-oxidation process in and of itself is expected to reduce the quantities to be stored on-site by more than one-third.

The pyro-oxydation of $CaCl_2$/NaCl base americium extraction salts leads to oxides PuO_2 and probably AmO_2, but the yield of concentration in the black salt is lower and the white salt cannot be discarded as LLW.

During vacuum distillation, excess carbonate can dissociate and damage the efficiency of the process. Appropriate chlorine sparging at the end of the oxidation can eliminate this carbonate.

Introduction

The purification of plutonium generates spent salts, which are temporarily stored in a nuclear facility. A development programme for a pyrochemical treatment is in progress to stabilise these waste salts and to reduce the quantities stored on-site. Inspired by work previously undertaken by LANL [1] the treatment consists of a pyro-oxidation of the bulk salt then of a vacuum distillation to separate the volatile salt matrix from actinides oxides. The final objective is to recover a plutonium and americium heel suitable for long-term storage and to discard the chloride salt as low level waste (LLW or A-type waste). So the salt product shall meet the LLW operational limit of 0.43 g per batch.

The objective of the pyro-oxidation is to convert all the actinides species present in the spent salts – chlorides, metal, oxychlorides – into oxides suitable for vacuum distillation. Sodium carbonate (Na_2CO_3), which is soluble in the molten solvent salt at 800°C, has been tested as an oxidising agent.

Pyro-oxidation reactions

Composition of the spent salt

The mean qualitative composition is given in Table 1. The variability of actinides concentration and nature is large. For electrorefining (ER) the metal fraction dispersed as droplets in the salt matrix can be important especially for salt containing few hundred grams of plutonium. In americium extraction (AE) salts the actinide chloride fraction is more important.

Reactions

The reactions have been settled by using a Gibbs energy minimisation program [2,3] for each plutonium species. The formation of plutonium oxides, as dioxide PuO_2 and sesquioxide αPu_2O_3, is expected. In the case of $PuCl_3$ an insufficient addition of carbonate can lead to the presence of undesirable oxychloride.

$$2Pu + Na_2CO_3 \rightarrow Pu_2O_3 + 2Na + C$$

$$2PuCl_3 + 3Na_2CO_3 \rightarrow 2PuO_2 + 6NaCl + CO + 2CO_2$$

$$2PuOCl + Na_2CO_3 \rightarrow 2PuO_2 + 2NaCl + CO$$

Experimental procedure

The spent salt and sodium carbonate are heated at 800°C in a magnesia crucible and stirred during four hours. Then stirring is stopped and the mixture is kept at 800°C for more than 1 hour for decantation of insoluble species. After a controlled cooling down to room temperature the reaction products are extracted from the crucible and separated.

The molar ratio Na_2CO_3/Pu is 1.5 for ER spent salts and 2 for AE spent salts.

The oxidations are performed in conventional plutonium pyrochemical processing glove boxes (GB) (see Figure 1).

Input and output products are weighed and Pu and Am are determined by non-destructive assays (calorimetry and gamma spectrometry). Output products are sampled for radiochemical analysis and X-ray diffraction (XRD) characterisation.

Results and discussion

NaCl/KCl base spent salts

For both ER and AE equimolar NaCl/KCl base spent salts, decantation of oxides results in a two-phases product. The bottom phase or black salt contains more than 97% of the actinides (see Tables 2 and 3). The top phase or white salt can be discarded as LLW.

Radiochemical analyses of samples taken in the core of the white salt show very low actinide concentrations: < 0.05 ppm for Am and 10 ppm for Pu. These results suggest that the residual actinides in the white salt are trapped at the edges and at the interface.

Samples of four ER spent salts oxidation products were analysed by XRD. The results are summarised in Table 4. In the black salt, PuO_2 is identified as the major plutonium species. Sodium plutonates, especially Na_4PuO_4, seem to be produced as well (Figure 2). Nevertheless, most of the rays attributed to sodium plutonates can be attributed to sub-oxides PuO_{2-x}. Thus the occurrence of Na_4PuO_4 is probable but needs to be confirmed. This uncommon crystal was previously identified by S. Pillon by contacting PuO_2 and Na_2O [4]. It is worth noting that neither Pu metal nor carbides are detected. However, according to the detection limit of XRD (few weight per cent), the occurrence of these compounds cannot be totally excluded.

Once the white salt is discarded, the quantities to store on-site are reduced by more than one-third. The MgO crucible can be reused for several successive treatments.

CaCl₂/NaCl spent salts

Few tests were carried out with equimolar $CaCl_2/NaCl$ AE salts. The reaction product is diphasic, and more than 80% of initial actinides are concentrated in the bottom black salt. The top white salt contains between 0.5-2 g of actinides. These actinides were in the form of particles spread in the salt matrix and as a vacuolar deposit at the top resulting from foam solidification. Otherwise, salts and particles projections adhere to the crucible walls.

Foam and projections suggest that gas exhaust is more important in the case of $CaCl_2/NaCl$. This could be explained by the distribution of the actinides species in the spent salt where the actinides chlorides and oxychlorides are predominant.

XRD analyses identify dioxides PuO_2 and probably AmO_2 as the lines of both are very close (see Table 5, Figure 3). Dedicated cold experiments have confirmed the presence of calcium carbonate resulting from the following reaction:

$$CaCl_2 + Na_2CO_3 \rightarrow CaCO_3 + 2NaCl$$

At 800°C the dissociation of pure calcium carbonate is significant and results in gas exhaust. However, calcium oxide is not detected by XRD.

$$CaCO_3 \rightarrow CaO + CO_2$$

Elimination of excess carbonate

According to the variability of the initial speciation of actinides, a large amount of sodium carbonate can remain in the products. Excess carbonate damages the efficiency of the vacuum distillation process due to dissociation and the presence of Na_2O and Na_2CO_3 in the final Pu heel.

A chlorine sparging sequence at the end of oxidation can eliminate this carbonate excess.

$$Na_2CO_3 + Cl_2 \rightarrow 2NaCl + CO_2 + 1/2\ O_2$$

Cold tests on $NaCl/KCl/Na_2CO_3$ have shown that stoichiometric chlorine addition leads to complete elimination. Detection of chlorine in exhaust gas marks the end of the reaction.

Argon/chlorine sparging has been tested on ER spent salt. This treatment does not alter the oxidation performances (Table 4) by limiting the chlorine addition. For the first two tests, the chlorine gas was detected in exhaust after an excess addition compared to the initial carbonate. The white salt does not meet LLW criterion and corrosion causes breakage of the metallic sparge tube.

The bottom product contains a large greenish-brown well-crystallised zone typical of PuO_2. At the edges and at the base some smaller black zones occur.

PuO_2 is the only detectable Pu species by XRD analysis in the first type zone. In the black zones PuO_2 and sodium plutonates (or PuO_{2-x}) are identified. On both samples sodium carbonate remains undetected.

Conclusion

Pyro-oxidation with sodium carbonate is very effective for NaCl/KCl spent salts in terms of both plutonium oxidation and waste reduction. The white salt can be discarded as low level waste (A-type waste) and oxidation by itself can reduce the quantities to be stored on-site by more than one-third. An appropriate chlorine sparging eliminates the carbonate in excess.

For $CaCl_2/NaCl$ americium extraction spent salts, the oxidation is effective but foaming damages the efficiency in term of waste reduction. The carbon oxides gas exhaust could be amplified by the nature of the solvent salt.

REFERENCES

[1] J.L. McNeese, "Pyrochemical Oxidation of Salt Residues", 21st Actinides Separation Conference, Charleston, SC, 23-26 June 97.

[2] THERMODATA – INPG – CNRS, thermochemical data bank COACH, Thermodata, 38402 St. Martin d'Hères.

[3] THERMODATA – INPG – CNRS, program GEMINI 1, Gibbs Energy Minimiser Thermodata, 38402 St.-Martin-d'Hères.

[4] S. Pillon, "Étude des Diagrammes de Phases U-O-Na, Pu-O-Na, U,Pu-O-Na", thesis 1989, Université des Sciences Techniques du Languedoc, Montpellier, France.

Table 1. Typical composition of plutonium purification spent salts (in weight %)

Origin	Nature	% Pu	% Am	Form
Extraction Am	NaCl/KCl CaCl$_2$/NaCl	2-15	< 2	Chlorides Oxychlorides Oxides
Electrorefining	NaCl/KCl	2-25	< 0.5	Metal ($\approx$50%) Oxychlorides Chlorides Oxides, Oxyhydrides

Table 2. Distribution of plutonium in ER and AE NaCl/KCl salts products (NDA assays: ± 6%)

	Feed salt Pu (wt.%)	Black salt Pu (wt.%)	White salt Pu (wt.%)	Waste type
ER/01	12.7	19.6	0.024	A
ER/02	11.9	26.0	0.012	A
ER/03	9.2	26.4	0.032	A
ER/04	18.0	27.6	0.034	A
ER/05	10.8	26.5	0.010	A
ER/06	10.6	16.4	0.003	A
ER/09	13.5	20.2	0.007	A
AE/01	1.8	4.1	0.038	A
AE/02	6.1	12.3	0.038	A

Table 3. Distribution of americium in ER and AE NaCl/KCl salts products

	Feed salt Am (ppm)	Black salt Am (ppm)	White salt Am (ppm)
ER/01	104 ± 21	–	< 0.05
ER/02	68 ± 14	300 ± 3	< 0.05
ER/03	9 ± 2	50 ± 5	< 0.05
ER/09	$1\,030 \pm 210$	$1\,120 \pm 220$	< 10
AE/01	$2\,210 \pm 660$	$4\,850 \pm 1\,450$	< 50
AE/02	$2\,700 \pm 810$	$5\,700 \pm 1\,710$	< 50

In italic: Determined by radioanalyses of samples.

Table 4. XRD characterisation results of ER pyro-oxidation products

White salt	Black salt
NaCl , KCl	PuO_2
$(Na_xK_{(1-x)})Cl$ solid solution	Na_4PuO_4, $PuO_{(2-x)}$
Na_2CO_3 traces	NaCl, KCl, Na_2CO_3 traces

Table 5. Distribution of Pu in ER pyro-oxidation/chlorination products (NDA assays: $\pm 6\%$)

	Cl_2/Na_2CO_3	Feed salt Pu (wt.%)	Black salt Pu (wt.%)	White salt Pu (wt.%)	Waste type
ER/07	1.5	11.6	24.5	0.107	B
ER/08	1.35	14.7	25.9	0.063	B
ER/10	0.66	13.7	26.2	0.006	A
ER/11	0.66	14.4	26.5	0.008	A
ER/12	1	4.2	9.3	0.008	A

Figure 1. Equipment for pyro-oxidation

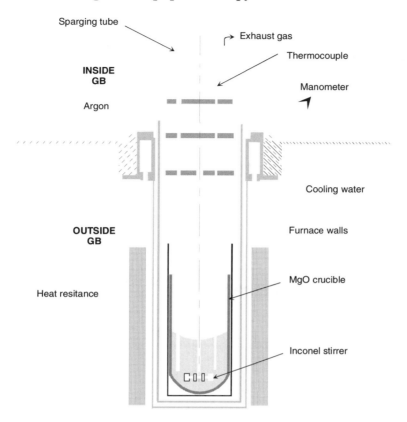

Figure 2. XR diffractogram of ER oxidation black salt

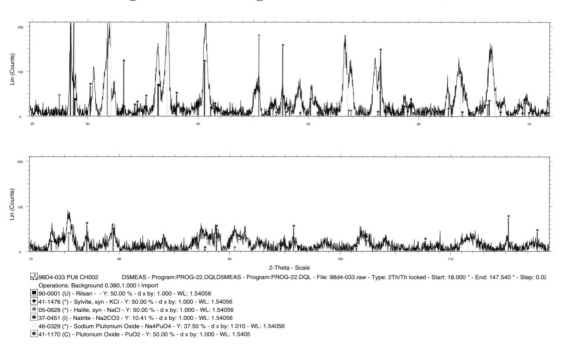

98D4-033 PU8 CH002   D5MEAS - Program:PROG-22.DQLD5MEAS - Program:PROG-22.DQL - File: 98d4-033.raw - Type: 2Th/Th locked - Start: 18.000 ° - End: 147.540 ° - Step: 0.02
Operations: Background 0.380,1.000 l Import
90-0001 (U) - Rilsan - - Y: 50.00 % - d x by: 1.000 - WL: 1.54056
41-1476 (*) - Sylvite, syn - KCl - Y: 50.00 % - d x by: 1.000 - WL: 1.54056
05-0628 (*) - Halite, syn - NaCl - Y: 50.00 % - d x by: 1.000 - WL: 1.54056
37-0451 (I) - Natrite - Na2CO3 - Y: 10.41 % - d x by: 1.000 - WL: 1.54056
46-0329 (*) - Sodium Plutonium Oxide - Na4PuO4 - Y: 37.50 % - d x by: 1.010 - WL: 1.54056
41-1170 (C) - Plutonium Oxide - PuO2 - Y: 50.00 % - d x by: 1.000 - WL: 1.5405

293

Figure 3. XR Diffractogram of CaCl₂/NaCl AE pyro-oxidation black salt

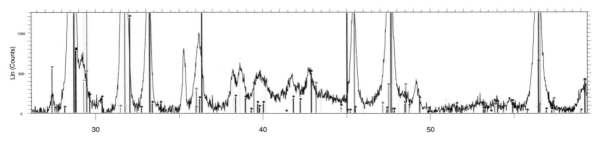

$PuO_2 - AmO_2 - CaCl_2 - CaCl_2,2H_2O - NaCl - CaCO_3$

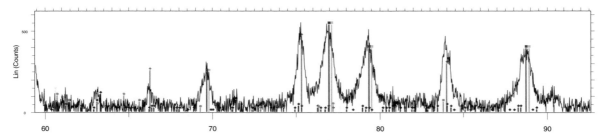

2-Theta - Scale

99D5-042 du 07/05/99   D5000 - File: 99d5-042.raw - Type: 2Th/Th locked - Start: 18.000 ° - End: 150.000 ° - Step: 0.020 ° - Step time: 10. s - Temp.: 25 °C (Room) - Time Started: 3 s - 2-Theta: 18.000 ° - Theta:
90-0001 (U) - Rilsan - - Y: 50.00 % - d x by: 1. - WL: 1.54056
75-2011 (C) - Plutonium Oxide - PuO2 - Y: 31.25 % - d x by: 1. - WL: 1.54056
75-2012 (C) - Americium Oxide - AmO2 - Y: 31.25 % - d x by: 1. - WL: 1.54056
05-0628 (*) - Halite, syn - NaCl - Y: 27.08 % - d x by: 1. - WL: 1.54056
70-0385 (C) - Calcium Chloride Hydrate - CaCl2(H2O)2 - Y: 7.64 % - d x by: 1. - WL: 1.54056
24-0223 (*) - Hydrophilite [NR] - CaCl2 - Y: 8.33 % - d x by: 1. - WL: 1.54056
05-0586 (*) - Calcite, syn - CaCO3 - Y: 12.50 % - d x by: 1. - WL: 1.54056

ELECTROCHEMICAL STUDIES OF EuCl₃ AND EuCl₂ IN AN EQUIMOLAR NaCl-KCl MELT

Sergey A. Kuznetsov

Institute of Chemistry Kola Science Centre RAS
184200 Apatity, Murmansk Region, Russia

Marcelle Gaune-Escard

IUSTI-CNRS UMR 6595, University of Provence
F-13453 Marseille Cedex 13, France

Abstract

The electroreduction/electro-oxidation process Eu(III) + e ⇔ Eu(II) in equimolar NaCl-KCl melt in the temperature range 973-1 123 K on glassy carbon electrode was studied using linear sweep voltammetry. It was determined that at a polarisation rate of $V \leq 0.1$ Vs⁻¹ the process of electroreduction of Eu(III) to Eu(II) is reversible, but at $0.1 < V \leq 0.3$ Vs⁻¹ a mixed diffusion and electron-transfer control was observed. The diffusion coefficients of Eu(III) and Eu(II) were determined by linear sweep voltammetry. The standard rate constants for the reaction of recharge of Eu(III) to Eu(II) were calculated on the basis of cyclic voltammetry data. The sluggish kinetics of this reaction is discussed in terms of substantial rearrangement of the europium co-ordination sphere. The formal standard potentials $E^*_{Eu(II)/Eu}$, $E^*_{Eu(III)/Eu}$ and formal redox potentials $E^*_{Eu(III)/Eu(II)}$ were determined from open-circuit potentiometry and linear sweep voltammetry data.

Introduction

The lanthanide elements which are present in spent fuel from fast nuclear reactors can be converted into molten salts by anodic dissolution [1]. The actinides are selectively deposited at the cathodes due to the differences among the redox potentials of the elements while fission products remain in the anode and in the electrolyte [1-2]. The lanthanide elements are the most difficult fission products to separate from actinides due to their similar chemical properties. Notwithstanding the active research, especially on rare earth metal electrowinning and electrorefining, the important fundamental problem of electrochemistry and thermodynamics of some rare earth metals in melts remains outside the field of vision.

Thus a knowledge of the electrochemistry and formal standard potentials of europium (III, II) in molten salts is very useful for the understanding the recycling of spent fuel.

To our knowledge no detailed electrochemistry and formal standard redox potentials of $E^*_{Eu(III)/Eu(II)}$ in equimolar NaCl-KCl melt have been reported before. Therefore in this article the electrochemistry of Eu(III), Eu(II), the formal standard potentials $E^*_{Eu(II)/Eu}$, $E^*_{Eu(III)/Eu}$ and formal standard redox potentials $E^*_{Eu(III)/Eu(II)}$ in NaCl-KCl melt are described.

Experimental

Europium dichloride was synthesised from the oxide Eu_2O_3 (Johnsen Matthey, 99.9%). Thionyl chloride (Jonhson Matthey, 99%) was used as a chlorinating agent during six hours at the temperature of 823 K under argon flow. $EuCl_3$ was obtained in the first step of this synthesis. Reduction to $EuCl_2$ was performed at 773 K during three hours by zinc under dynamic vacuum. Then the temperature was increased to 1 093 K and maintained constant for five hours under static vacuum. Finally, the $EuCl_2$ compound was purified from zinc by distillation at 1 193 K. Due to the highly hygroscopic properties of lanthanide compounds, $EuCl_3$ and $EuCl_2$ were stored in sealed glass ampoules under vacuum. All further handling of europium chlorides and filling of experimental cells were performed in a controlled purified argon atmosphere glove box containing less than 2 ppm of water.

Alkali chlorides (NaCl and KCl) were purchased from Prolabo (99.5% min.). They were dehydrated by continuous and progressive heating just above the melting point under gaseous HCl atmosphere in quartz ampoules. Excess HCl was removed from the melt by argon. The salts were handled in the glove box and stored in sealed glass ampoules, as explained above. Chlorides of sodium and potassium were mixed in a required ratio, placed in an ampoule fabricated of glassy carbon of the SU-2000 type and transferred to a hermetically sealed retort of stainless steel. The latter was evacuated to a residual pressure of $5 \cdot 10^{-3}$ Torr, first at room temperature and then gradually heating to 473, 673 and 873 K. Afterwards the retort was filled with high purity argon and the electrolyte was melted.

The study was performed employing the linear sweep voltammetry (LSV) using a VoltaLab-40 potentiostat with packaged software "VoltaMaster 4". The potential scan rate was varied between $5 \cdot 10^{-3}$ and 5.0 Vs^{-1}. The experiments were carried out in a temperature range 973-1 123 K. The cyclic voltammetric curves were recorded at 0.8-2.0 mm diameter glassy carbon and platinum electrodes with respect to a glassy carbon plate as a quasi-reference electrode and to a silver reference electrode, Ag/NaCl-KCl-AgCl (2 wt.%). The glassy carbon ampoule served as the counter electrode. The potentials from the silver reference electrode were converted to a Cl^-/Cl_2 reference electrode using the relation:

$$E/V = -1.111 - 1.31 \cdot 10^{-4} \ T/K \qquad (1)$$

Results and discussion

Cyclic voltammetry of the electrode reaction Eu(III) + e ⇔ Eu(II)

The cyclic voltammetric curves in the NaCl-KCl-EuCl₃ melt obtained at the glassy carbon electrode are presented in Figure 1. Wave 1 is observed in the cathodic-anodic region that indicates an appearance in the melt of Eu(II) due to the reaction:

$$2EuCl_3 \Leftrightarrow 2EuCl_2 + Cl_2 \tag{2}$$

Thisis not a too unexpected result because it is known from [3] that the trichloride of europium starts to decompose into dichloride and chlorine in solid phase at temperatures above 300°C.

The potentiostatic electrolysis at potentials of the cathodic peak did not lead to the formation of a solid phase at the electrode and the electrode itself underwent no visible transformation. This means that the product at this stage is soluble in the melt. We investigated the dependencies of the electroreduction peak current and the peak potential on the polarisation rate (Figures 2 and 3). It was found that the peak current is directly proportional to the square root of the polarisation rate (Figure 3a), while the peak potential does not depend on the polarisation rate up to $V = 0.1$ Vs^{-1} (Figure 3b). The peak current of the electroreduction process is linearly dependent on the EuCl₃ concentration, while the peak potential of the first stage does not depend on the concentration of europium trichloride in the melt. According to the theory of linear sweep voltammetry [4], up to a polarisation rate of 0.1 Vs^{-1} the electrode process is controlled by the rate of mass transfer and yields a reduced form soluble in the melt. The number of electrons for the first stage was calculated by means of the following equations [5]:

$$E_{p/2}^C - E_p^C = 2.2\,RT/nF \tag{3}$$

$$E_p^A - E_p^C = 2.2\,RT/nF \tag{4}$$

where E_p^C is the potential of the cathodic peak, $E_{p/2}^C$ is the potential of the half-peak, n is the number of electrons and E_p^A is the potential of the anodic peak.

The calculation of the number electrons revealed that one electron is transferred at the first stage:

$$Eu(III) + e \Leftrightarrow Eu(II) \tag{5}$$

The diffusion coefficients (D) for the chloride complexes of Eu(III) were determined at $V = 0.1$Vs^{-1} using of the Randles-Shevchik equation [5]:

$$I_p^C = 0.4463F^{3/2}R^{-1/2}T^{-1/2}n^{3/2}ACD^{1/2} \tag{6}$$

where I_p^C is the peak cathodic current (A), A is the electrode area (cm²), C is the bulk concentration of active species (molcm^{-3}), D is the diffusion coefficient (cm²s^{-1}), V is the potential sweep rate (Vs^{-1}) and n is the number of electrons involved in the reaction.

The coefficients are described by the following empirical dependence:

$$\log D_{Eu(III)} = -2.42 - 2152/T \pm 0.03 \qquad (7)$$

Eq. (6) was employed for calculation of $D_{Eu(II)}$ in the melt NaCl-KCl-EuCl$_2$ on the basis of the peak current determined for the process:

$$Eu(II) - e \rightarrow Eu(III) \qquad (8)$$

At polarisation rate 0.1 Vs^{-1} in the temperature range 973-1 123 K the experimental values fit a straight line in a log D vs. 1/T plot. The straight line can be described by the following empirical equation:

$$\log D_{Eu(II)} = -2.31 - 1983/T \pm 0.03 \qquad (9)$$

Determination of the standard rate constants of the electrode reaction Eu(III) + e ⇔ Eu(II)

Analysing the dependencies presented in Figure 3 shows that, at $0.1 < V \leq 0.3$ Vs^{-1}, the process in Eq. (5) is quasi-reversible. A mixed diffusion and electron-transfer rate controlled the process (5) at $V > 0.1 Vs^{-1}$ is indicated by the deviation of the experimental points from straight line in the I_p^C vs. $V^{1/2}$ plots (Figure 3a) by the dependence E_p^C on V (Figure 3b), and by the magnitude of the difference between E_p^A and E_p^C, which is larger than is required for a reversible process. For example, at a polarisation rate of 0.2 Vs^{-1} it is equal to 0.338 V. A further increase of polarisation rate $V \geq 0.5$ Vs^{-1} results in electron-transfer control.

The problem of the determination of kinetic parameters on the basis of cyclic voltammetry was considered by Nicholson [6]. The standard rate constant of the electrode process is related to the function ψ as follows:

$$\Psi = \frac{k_s \left(\dfrac{D_{ox}}{D_{red}} \right)^{\alpha/2}}{\pi^{1/2} D_{ox}^{1/2} \left(\dfrac{nF}{RT} \right)^{1/2} V^{1/2}} \qquad (10)$$

Here Ψ is function related to the difference between the peaks potentials $E_p^A - E_p^C$ (mV), k_s is the standard rate constant of electrode process (cms^{-1}) and $\alpha = 0.5$ is the transfer coefficient.

The dependencies $E_p^A - E_p^C$ on the function Ψ reported in [6] for a temperature of 298 K, should be recalculated for the working temperature. The recalculation was performed using the following equations [7]:

$$\left(\Delta E_p \right)_{298} = \left(\Delta E_p \right)_T 298/T \qquad (11)$$

$$\Psi_T = \Psi_{298} (T/298)^{1/2} \qquad (12)$$

The values of the ψ_T function, obtained by means of Eqs. (11) and (12) and used in conjunction with Eq. (10) made it possible to calculate the standard rate for charge transfer. As the temperature is elevated the standard rate constants grow and its values are equal to $0.80 \cdot 10^{-2}$, $1.25 \cdot 10^{-2}$ and $2.27 \cdot 10^{-2}$ cm^{2}s^{-1} at 973, 1 023 and 1 073 K respectively. The values of the constants testify that, at $V = 0.2$ Vs^{-1}, the process in Eq. (5) proceeds quasi-reversibly, mostly under diffusion control.

Formal standard redox potentials $E^*_{Eu(III)/Eu(II)}$

According to the theory of linear sweep and steady state voltammetry the following relation is valid for the reversible electrochemical reduction (5) between the cathodic and anodic peak potentials and half-wave potential [8]:

$$E_p^c = E_{1/2} - 1.11(RT/F) \tag{13}$$

$$E_p^a = E_{1/2} + 1.11(RT/F) \tag{14}$$

$$\left(E_p^c + E_p^a\right)/2 = E_{1/2} \tag{15}$$

where:

$$E_{1/2} = E^0_{Eu(III)/Eu(II)} + RT/F \ln\left(D_{red}/D_{ox}\right)^{1/2} + RT/F \ln\left(\gamma_{ox}/\gamma_{red}\right) \tag{16}$$

In the concentration range of ions with mole fraction less than $(3-5) \cdot 10^{-2}$ the coefficients of activity in the molten salts remain constant leads up to the values of the formal standard potentials [9]:

$$E^*_{Eu(III)/Eu(II)} = E^0_{Eu(III)/Eu(II)} + RT/F \ln\left(\gamma_{ox}/\gamma_{red}\right) \tag{17}$$

Thus the formal standard redox potentials of $E^*_{Eu(III)/Eu(II)}$ can be calculated using the following:

$$E^*_{Eu(III)/Eu(II)} = E_{1/2} + RT\big/F \ln\left(D_{ox}/D_{red}\right)^{1/2} \tag{18}$$

$$E^*_{Eu(III)/Eu(II)} = E_{\tau/4} + RT\big/F \ln\left(D_{ox}/D_{red}\right)^{1/2} \tag{19}$$

$$E^*_{Eu(III)/Eu(II)} = E_p^C + 1.11(RT/F) + RT/F \ln\left(D_{ox}/D_{red}\right)^{1/2} \tag{20}$$

$$E^*_{Eu(III)/Eu(II)} = E_p^A + 1.11(RT/F) + RT/F \ln\left(D_{ox}/D_{red}\right)^{1/2} \tag{21}$$

$$E^*_{Eu(III)/Eu(II)} = \left(E_p^C + E_p^A\right)/2 + RT/F \ln\left(D_{ox}/D_{red}\right)^{1/2} \tag{22}$$

$$E^*_{Eu(III)/Eu(II)} = E_p + RT\big/\alpha n_\alpha F\left[0.78 - \ln k_s + \ln\left(\alpha n_\alpha F V D_{ox}/RT\right)^{1/2}\right] \tag{23}$$

For calculation of $E^*_{Eu(III)/Eu(II)}$ we used Eqs. (20-22) because the values E_p^C and E_p^A had a better reproducibility by comparison with values $E_{1/2}$ and $E_{\tau/4}$. Eq. (23) is valid for irreversible process and application of this expression includes a mistake in determination of the standard rate constant (k_s).

Thus using the values of potential peaks of the process (5) and the coefficients of diffusion of Eu(III) and Eu(II) it was found that the formal standard redox potentials are described by the following empirical dependence:

$$E^*_{Eu(III)/Eu(II)}/V = -(0.971 \pm 0.006) + (1.9 \pm 0.2) \cdot 10^{-4} T/K \tag{24}$$

Cyclic voltammetry of the electrode reaction Eu(II) + e ⇔ Eu

Using cyclic voltammetry it is not possible to determine the value of the electroreduction peak for:

$$Eu(II) + 2e \rightarrow Eu \tag{25}$$

because the potentials of discharge processes (25) and alkali metals are very similar. In the cathodic cycle of the voltammograms, the ascending sections at highly negative potentials resulted in simultaneous electroreduction of europium and alkali metals. The cyclic voltammetric curves in the NaCl-KCl-EuCl$_2$ melt on the different substrates from molybdenum, glassy carbon, platinum, copper, kanthal and silver were obtained. It was determined that the most suitable electrode for studying Eq. (25) is molybdenum because it does not form intermetallic compounds with europium (Figure 4). It is necessary to note that Eq. (5) can be studied on glassy carbon and platinum electrodes. Other electrodes which were used in this investigation have more negative potentials of dissolution than the potentials of electroreduction/electro-oxidation Eu(III) + e ⇔ Eu(II). For instance in the NaCl-KCl melt containing EuCl$_3$ molybdenum electrode interacted with the melt due to the reaction:

$$2Eu(III) + Mo \Leftrightarrow 2Eu(II) + Mo(II) \tag{26}$$

Thus cyclic voltammetry does not provide information on determination of formal standard potentials of $E^*_{Eu(II)/Eu}$. Therefore for these determinations we used an open-circuit potentiometry method.

Open-circuit potentiometry

Pure europium deposits are formed on a molybdenum electrode in the melt NaCl-KCl-EuCl$_2$ by constant current electrolysis with current densities 50-600 mA/cm^2 during 5-120 seconds. A typical potential decay obtained after electrolysis is shown in Figure 5. The length of potential plateau associated with dissolution of europium increased with increasing current densities and time of electrolysis. In some cases the potential plateau remained stable for more than 2-3 min. (Figure 5), then slowly shifted toward the positive potential and finally arrived at the rest potential of molybdenum in the melt. As can be seen from Figures 4 and 5 the values of potential plateau and potential which intersects cyclic voltammograms at zero current are identical. Thus in some cases the rough determination of formal standard potentials can be obtained by extrapolation of the ascending section of the voltammetric curve to the potential axis. Plots of the plateau potential against log N give linear relationships with slopes which are very close to the Nernst slope 2.3RT/2F for a reversible transfer of two electrons. A least squares calculation gives values of formal standard potentials $E^*_{Eu(II)/Eu}$ (referred to Cl$^-$/Cl$_2$ reference electrode) at each temperature, and the dependence on temperature is accurately represented by the empirical relation:

$$E^*_{Eu(II)/Eu}/V = -(4.08 \pm 0.01) + (8.2 \pm 0.2) \cdot 10^{-4} T/K \tag{27}$$

The formal standard potential $E^*_{Eu(III)/Eu}$ was calculated from Luter's equation:

$$3E^*_{Eu(III)/Eu} = E^*_{Eu(III)/Eu(II)} + 2E^*_{Eu(II)/Eu} \qquad (28)$$

and its temperature dependence is:

$$E^*_{Eu(III)/Eu} / V = -(3.044 \pm 0.009) + (6.1 \pm 0.2) \cdot 10^{-4} \, T/K \qquad (29)$$

Comparison between present results and calorimetric data

The change in the partial Gibbs energies for the formation of europium chlorides from the elements in NaCl-KCl were calculated using Eqs. (27) and (29). For the formation of europium chlorides in molten NaCl-KCl, the following changes in the partial Gibbs energies were obtained:

$$\Delta G^*_{EuCl_2} / kJmol^{-1} = -(787 \pm 2) + (158 \pm 4) \cdot 10^{-3} \, T/K \qquad (30)$$

$$\Delta G^*_{EuCl_3} / kJmol^{-1} = -(881 \pm 3) + (176 \pm 6) \cdot 10^{-3} \, T/K \qquad (31)$$

The literature data [10,11] on the thermodynamics of the reaction of liquid europium with chlorine with formation of dichloride and trichloride europium allows us to calculate the changes in standard enthalpy, which at 1 100 K are -789 kJmol^{-1} for EuCl$_2$ and -865 kJmol^{-1} for EuCl$_3$. Thus the change in the relative partial enthalpy of mixing EuCl$_2$ with equimolar mixture NaCl-KCl when dilute solutions are formed is, within experimental error, close to zero. This value is in good agreement with the partial enthalpies derived from integral enthalpies for the systems EuCl$_2$-NaCl and EuCl$_2$-KCl obtained by a calorimetric method [12]. Thus in the melt NaCl-KCl-EuCl$_2$, it appears that chloride complexes with Eu^{2+} are not significant. At the same time the partial enthalpy of mixing EuCl$_3$ with NaCl-KCl melt at the temperature 1 100 K is -16 kJmol^{-1} due to the complex formation reaction:

$$Eu^{3+} + 6Cl^- \Leftrightarrow EuCl_6^{3-} \qquad (32)$$

Peculiarity of intervalence charge transfer of the couple Eu(III)/Eu(II)

The peculiarity of redox electrochemistry of the couple Eu(III)/Eu(II) is a sluggish kinetics of the reaction in Eq. (5). The electron transfer itself, which in the condensed phase may occur on a time scale of 10^{-15} s, is preceded by the slower step of complex rearrangement [13]. The energy and the time required for the complex rearrangement decreases as the scale of chemical bond rearrangement decreases. Related with this is the fact that complex recharge usually is a reversible process (even at high polarisation rates up 5-10 Vs^{-1}) [14], because these reactions can in principle occur only with a change of charge and without a change in the composition of the complexes. The sluggish kinetics of the reaction:

$$EuCl_6^{3-} + e \rightarrow Eu^{2+} + 6Cl^- \qquad (32)$$

is connected with substantial rearrangement of the europium co-ordination sphere (probably the loss of six ligands) which occurs during the electrode reaction.

Conclusion

The electrochemical reduction of Eu(III) in NaCl-KCl equimolar mixture occurs via two successive reversible stages involving the transfer of one and two electrons. The diffusion coefficients of Eu(III) and Eu(II) were determined by linear sweep voltammetry. The standard rate constants for the reaction of the intervalence charge transfer were calculated on the basis of cyclic voltammetry data. The possibility of the determination of partial Gibbs energy during the mixing of europium trichloride and dichloride in NaCl-KCl melt using electrochemical transient techniques was shown. The formal standard potentials $E^*_{Eu(II)/Eu}$, $E^*_{Eu(III)/Eu}$ and formal redox potentials $E^*_{Eu(III)/Eu(II)}$ are determined from open-circuit potentiometry and linear sweep voltammetry data. The thermodynamics of formation of dilute solutions of europium di- and trichloride in NaCl-KCl melt was calculated. The value for the partial enthalpy of mixing $EuCl_2$ in NaCl-KCl from electrochemical measurements is in good agreement with data obtained using the calorimetric method.

REFERENCES

[1] T. Koyama, M. Iizuka, N. Kondo, R. Fujita and H. Tanaka, *J. Nucl. Mater.*, 247 (1997), 227.

[2] M. Iizuka, T. Koyama, N. Kondo, R. Fujuita and H. Tanaka, *J. Nucl. Mater.*, 247 (1997), 183.

[3] D.M. Laptev, T.V. Kiseleva, N.M. Kulagin, *Russ. J. Inorg. Chem.*, 31 (1986), 1965.

[4] R.S. Nicholson and I. Shain, *Anal. Chem.*, 36 (1964), 706.

[5] Z. Galus, "Fundamentals of Electrochemical Analysis", Ellis Horwood, London (1994).

[6] R.S. Nicholson, *Anal. Chem.* 37 (1965), 1351.

[7] S.A. Kuznetsov, S.V. Kuznetsova and P.T. Stangrit, *Soviet Electrochemistry*, 26 (1990), 55.

[8] H. Matsuda and Y. Ayabe, *Z. fur Elektrochemie*, 59 (1955), 494.

[9] M.V. Smirnov, "Electrode Potentials in Molten Chlorides", Moscow, Nauka (1973).

[10] C.E. Wicks and F.E. Block, "Thermodynamic Properties of 65 Elements – Their Oxides, Halides, Carides and Nitrides", Bureau of Mines, Washington (1963).

[11] L.B. Pankratz, "Thermodynamic Properties of Halides", Bureau of Mines, Washington (1984).

[12] M. Gaune-Escard, F. Da Silva, L. Rycerz, *et al.*, in preparation.

[13] L.I. Krishtalik, "Electrode Reactions. The Mechanism of the Elementary Act", Moscow, Nauka (1982).

[14] S.A. Kuznetsov, *Russ. J. Electrochemistry*, 29 (1993), 1154.

Figure 1. Cyclic voltammetric curves at a glassy carbon electrode in NaCl-KCl-EuCl₃ melt

Area: 0.18 cm², Sweep rate: 0.15 Vs⁻¹, Temperature: 973 K,
Concentration of EuCl₃: 6.64·10⁻⁵ molcm⁻³, Reference electrode: Cl₂/Cl⁻

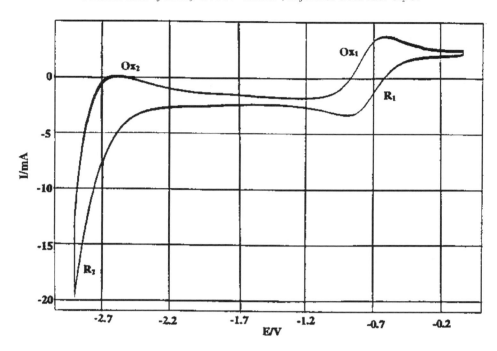

Figure 2. A series of cyclic voltammograms at a glassy carbon electrode for various scan rates in NaCl-KCl-EuCl₃ melt

Area: 0.19 cm², Temperature: 973 K,
Concentration of EuCl₃: 9.0·10⁻⁵ molcm⁻³, Reference electrode: Cl₂/Cl⁻
1 - V = 0.05 Vs⁻¹; 2 - V = 0.075 Vs⁻¹; 3 - V = 0.01 Vs⁻¹

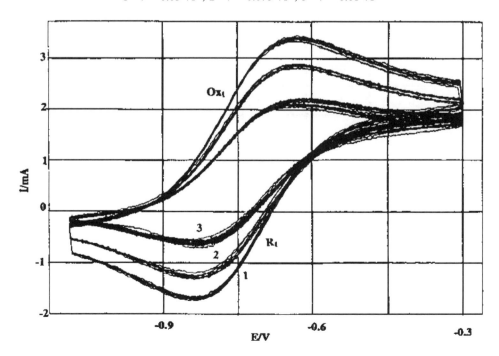

Figure 3. The dependencies of peak currents (a) and peak potentials (b) of the recharge process on the polarisation rate

Area: 0.19 cm², Temperature: 973 K,
Concentration of EuCl₃: 9.0·10⁻⁵ molcm⁻³, Reference electrode: Cl₂/Cl⁻

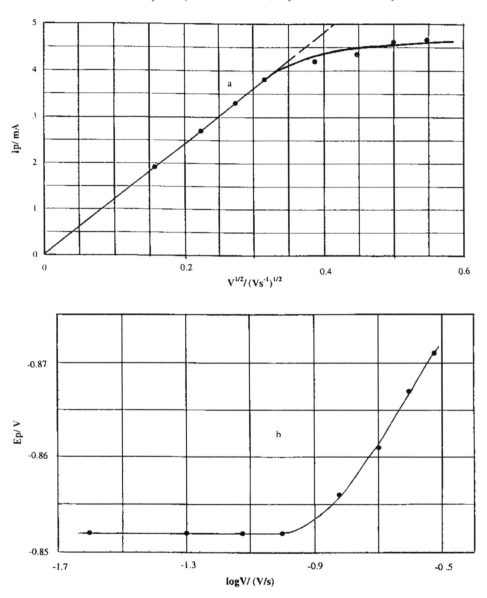

Figure 4. Cyclic volammetric curves recorded at a molybdenum electrode

Area: 1.0 cm², Temperature: 1 000 K, Sweep rate: 0.1 Vs⁻¹, Concentration of EuCl₃: 8.6·10⁻⁴ molcm⁻³,
Reference electrode: Cl₂/Cl⁻, Values of decay at a reverse potential: 1-0 sec., 2-1 sec., 3-2 sec.

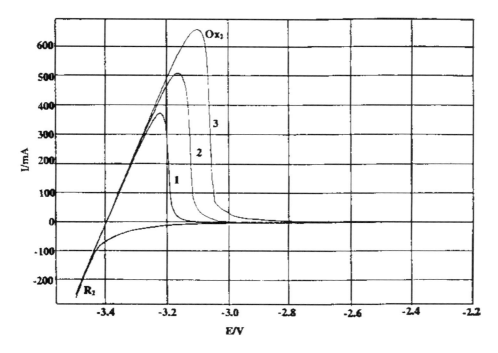

Figure 5. Open-circuit potential decay curves of molybdenum electrode after polarisation

Temperature: 1 000 K, Current density: 360 mAcm⁻²,
Concentration of EuCl₃: 8.6·10⁻⁴ molcm⁻³, Reference electrode: Cl₂/Cl⁻, Time of polarisation: 90 sec.

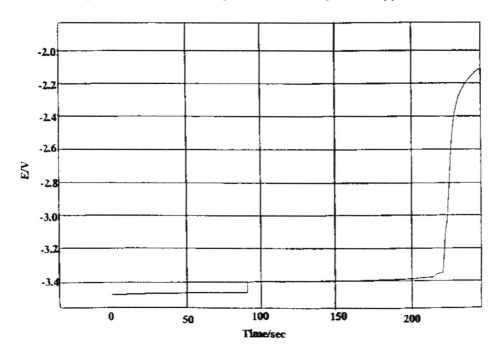

GENERALISED PERTURBATION THEORY AND SOURCE OF INFORMATION THROUGH CHEMICAL MEASUREMENTS

Vladimír Lelek, Tomáš Marek
Nuclear Research Institute Řež, Czech Republic

Abstract

It is important to make all analyses and collect all information from the work of the new facility (which the tranmutational demonstrational unit will surely be) to be sure that the operation corresponds to the forecast or to correct the equations of the facility. The behaviour of the molten salt reactor and in particular the system of measurement are very different from that of the solid fuel reactor. Key information from the long time kinetics could be the nearly on line knowledge of the fuel composition. In this work it is shown how to include it into the control and use such data for the correction of neutron cross-sections for the high actinides or other characteristics. Also the problem of safety – change of the boundary problem to the initial problem – is mentioned.

The problem is transformed into the generalised perturbation theory in which the adjoint function is obtained through the solution of the equations with right hand side having the form of source. Such an approach should be a theoretical base for the calculation of the sensitivity coefficients.

Introduction

There is a remarkable difference from the point of view of the flow of information between the liquid and solid fuel reactors. It is possible to have practically on line fuel composition during burn-up, and due to the possibility of the continuous withdrawal of some gases (such as Xe) from the primary circuit we suppose that the yield of some fission products will also be changed. On the other hand, the flow time of the fuel particles through the reactor is much smaller than the burn-up time, and we can suppose that the liquid fuel is homogenous and does not depend on its placement in the primary circuit. Naturally in such a study we neglect the effects arising from the temperature changes within the core and use mean temperature only. (To take into account temperature changes we must solve another class of problems including delayed neutrons.)

During the past several decades only solid fuel reactors were built up and therefore software was designed only for that type of reactor. The main difference is that each piece of fuel has its own burn-up and fission products which become progressively different with rising burn-up, and as a result it is necessary to take into account the space distribution of materials. The non-linearity of equations and the difficulty in independently solving each piece of the reactor are solved via approximate formulae for each space element, but the explicit expressions for the vector of fuel and fission products concentration is missing. Such an approach was also taken in the work [6,7], in which mathematical formalism on how to calculate indeterminateness in the reactor characteristics due to nuclear data was studied.

We want to give more explicit expressions and formulations on how to proceed during the burn-up calculation to be able to use the huge amount of chemical information about the fuel composition for a better control of the reactor and also for the correction of nuclear data. Which isotopes should be measured and how to interpret those measurements is a particular problem also connected with the fission yields (indicative and cumulative) and how detailed the burn-up equation is written. It is also different from the solid fuel reactors, in which the chemical measurements of the spent fuel are realised only after the fuel has been withdrawn from the reactor and had some cooling time.

The mathematical concept is to write equations with an explicit form of burn-up description so as to be able to explicitly study perturbations due to cross-sections and concentration changes.

In straight analogy to [4,6,7] we can write:

$$-\nabla D_k \nabla \phi_k + \Sigma_k \phi_k = \sum_{l=1}^{k-1} \Sigma_{k,l} \phi_k + \lambda_k \sum_{l=1}^{g} \Sigma_l^f \phi_l + S_k \tag{1}$$

where $\Sigma = \sum_{i=1}^{I} N_i \sigma_i$ with i denoting fission product or fuel nuclei, S_k being the external source of neutron (if it exists – it could also be zero), and D_k being the diffusion coefficient.

There is no specific problem in writing the explicit form of D_k in $\sigma_{i,k}^{(a)}$, $\sigma_i^{(s)}$ and concentrations N_i, but the influence of burn-up is much smaller, main is the basic media like LiF+BeF$_2$. Those are the reasons why it should not be taken into account just now.

$$\int dV \sum_{k=1}^{g} Q_k \phi_k = P_w \tag{2}$$

308

$$\int dV \frac{dN_i}{dt} = -N_i \int dV \sum_k \sigma_{i,k} \phi_k - \lambda_i \int dV N_i + \int dV \sum_j \sigma^j_{i,k} \phi_k N_j + \eta_i \qquad (3)$$

where $\int dV$ denotes integration over the whole primary circuit (if it is denoted), V_c is core volume, V_o is out of core volume ($\phi \approx 0$ in V_o), and η_i is external burn-up compensation. Eq. (3) represents burn-up and material conservation law during the work of the MS reactor.

Remark 1

If we attempt to decipher feedback in the molten salt (MS) reactor we see that the primary circuit volume is not constant. It grows with the growing power and temperature, corresponding to the molten salt extension and due to η, the amount of mass slightly growing.

Strictly speaking the reactor vessel and tubes are also extended but the effect on reactivity is about a magnitude lower than the salt thermal extension.

Remark 2

Because the flow of MS is much quicker than the burn-up changes we can suppose that the MS is ideally mixed (constant in the whole circuit). It follows from this the Ni functions only as time and not as space variables. This is a great simplification in comparison with the solid fuel reactor where all variables must be taken in to account.

The standard form of η_i will be Pu + MA from the spent fuel (in this work constant isotopic vector C_i) so that:

$$\eta_i = \eta C_i \qquad (4)$$

Eqs. (1), (2) and (3) define our problem, which in other words denotes to find for the initial concentration of fuel ($t = 0$) (non-zero elements of the vector C_i). Then, during the time $0 \leq t \leq T$, the velocity of adding of fuel to compensate burn-up in such a way as to fulfil the condition in Eq. (2) – given power.

We know from physical reasons and experience that such a mathematical problem has a solution and that the parameter, which will be determined, is temperature which must have the so-called negative reactivity coefficient. This means that if we have higher temperature we have smaller flux or that if we add or change something in the reactor with the result that the flux rises, there will be compensation of this effect (due to the decreasing of σ or greater leakage from the reactor) which will lead to the stabilisation of flux on the higher temperature.

In the framework of such tasks we are asking about the sensitivity coefficients of N and η on the not very well known parameters such as $\sigma_{i,k}$ or on densities (which depend on temperatures). We should find the development of the measured values into the series of small changes of parameters. We also want to use the new experimental knowledge to correct them. It is a very complicated mathematical problem, which is not very frequently solved – we need something like a perturbation theory for the non-linear reactor equations, and this has not yet been developed for use.

We shall not go into the details (which will be presented in a special work) we shall only write down corresponding functionals and describe how the algorithms for the solution will be derived.

Let us formally denote (1), (2) and (3) as:

$$M(N, p_w, \sigma)\phi = 0 \tag{5}$$

$$Q(\phi, N, \sigma) - P_w = 0 \tag{6}$$

$$A(\phi, p_w, \sigma)N - \eta C = 0 \tag{7}$$

Remark 3

We do not write explicitly temperature but it should be mentioned that there is a direct connection between power and mean temperature, supposing that the heat exchanger function is constant.

Optimal description

We can now formulate several problems and show how they should be solved.

Fuel concentration and its regulations

The problem we want to study is a precise prediction of the fuel concentration and velocity of the external fuel adding η. It is clear that we need dependence of η on various σ (basic nuclear cross-sections). It could be supposed that we could repeatedly solve the problem with small changes of σ and obtain the difference from the subtraction of results. Such an approach leads to the subtraction of the nearly the same numbers, and this is generally not recommendable. We show that sensitivity of η to the error of σ can be expressed with the help of the generalised adjoint function $\Phi^* = (\phi^*, N^*)$ (see Appendix A, Eq. (A3)):

$$\frac{\partial \eta}{\partial \sigma} = \frac{N^+ \dfrac{\partial A}{\partial \sigma} N + \phi^+ \dfrac{\partial M}{\partial \sigma} \phi + \omega \dfrac{\partial Q}{\partial \sigma}}{N^+ C} \tag{8}$$

where N^+C has the meaning of the standard scalar product – integration over space and time.

Here we should like to note that we can expect that $\Phi^* = (\phi^*, N^*)$ could be solved in exactly the same manner as Eqs. (1)-(3), and that due to negative feedback it can represent an analogy of the subcritical operator with external source. We should like to emphasise that in Eq. (A3) all derivatives such as $\partial M/\partial \sigma$ make good sense due to the explicit form of burn-up equations.

Use of core measurement

The second group of problem is formulated in the following question: Are we able to use measurements of core fluxes and especially the core composition to calculate or correct the errors in σ

or the influence of other parameters which will not be known at the beginning of the demonstration facility or not properly represented in our descriptions? We can follow Eq. (6) and write the following functional for measurement:

$$J = J_\sigma + J_\gamma \qquad (9)$$

where $J_\sigma = \sum_{ij} w_{ij}(\sigma_i - \overline{\sigma}_i)(\sigma_j - \overline{\sigma}_j)$ is positive definite expert estimation or errors of the nuclear data from the data files. The parameters $\overline{\sigma}_i, \overline{\sigma}_j$, are initial estimations.

$$J = J_\sigma + J_r(C_o, \gamma, C_e, \rho, \phi, \phi^+) = \qquad (10)$$

$$J_\sigma + w_{c_0}(C_0 - \overline{C}_0)^2 + \sum_\rho w_{c_0,\rho}\left(\frac{\partial C_0}{\partial \rho} - \frac{\partial \overline{C}_0}{\partial \rho}\right)^2 +$$

$$\int_0^t w_\eta[\eta(t') - \overline{\eta}(t')]^2 dt' + \sum_\rho \int_0^t w_{\eta,\rho}\left(\frac{\partial \eta}{\partial \rho} - \frac{\partial \overline{\eta}}{\partial \rho}\right)^2 dt' +$$

$$\sum_{det} w_{det}\left(\Sigma_{det}\phi - \overline{\Sigma_{det}\phi}\right)^2 + \sum_i \int w_i\left(N_i - \overline{N_i}\right)^2 dt$$

where ρ could be external parameters used for the time dependent measurements or also methodical parameters of the chemical and physical measurements.

The last term is the main source of information; it is practically full of information about time dependent concentration of all elements we shall decide or we shall be able to measure. Our problem is to use the uncertainties in σ to find the best description of the experimental data – this means to minimise J in the space of σ. In Appendix B the method of generalised Lagrange multipliers is suggested, which could be used to find unknown parameters and analyse the problem. The authors have practical experience in solving this type of problem.

Safety problems

If we omit in the Eq. (A2) the condition for power (proportional to ω) we shall have a transitional process – a mathematical task with initial conditions. The system, under external change of the chosen parameters, will develop alone and due to feedback will be stabilised in the new stage. Our problem is to show that the transitional process is sufficiently smooth and without additional maxims or minims is limited to the new asymptotic stage.

Problem formulation through the minimisation of the non-linear functional gives us the possibility to make more thorough analyses and study the influence of a broad class of parameters.

Conclusions

It is probably much easier to develop a new mathematical approach than to complete full knowledge of nuclear data for the new group of elements.

It is a special feature of the molten salt reactor that we shall be able to obtain a huge amount of information on-line through chemical measurement, and will pay for it only by slightly smaller physical measurement of fluxes. It is also important that contemporary computer techniques enable us to solve new classes of tasks in the "on-line" time. This situation should be reflected in the preparation of a new algorithm enabling us deeper and more precise understanding of the ongoing processes in the reactor including the "on-line" use of the reactor as the source of information for the better prediction of the operated system. It is in all cases better to let computers work instead of heavy experimental facilities.

These are the reasons why we are convinced that it will be efficient to develop new mathematical approaches for the operation of the reactor for the end of nuclear fuel cycle, representing continuous evaluation of all information. In this way we shall have not only a model for the description of processes but will have the added advantage of experimental data. As with all new processes, however, it requires a lot of work.

REFERENCES

[1]   "ENDF-6 Formats Manual", IAEA-NDS-76 (1992).

[2]   V. Lelek, "Modeling of Old Ra-Pollution Diffusion with Nearly Unknown Source", 2nd Pugwash Workshop on the Future of the Nuclear Weapon Complexes of the FSU and the USA, Moscow, 20-23 Feb.1995.

[3]   J.L. Lions, "Contrôle Optimal de Systèmes Gouvernés par des Equations aux Dérivées Particules", Dunod, Paris (1968).

[4]   V. Lelek, M. Pecka, L. Vrba, "Correction of Equations Based on Measurements and Application to Theory and Experiment Analysis", IAEA-TECDOC-678, December 1992, IAEA Proc. of Tech. Com. Meeting and Workshop, Řež, Czechoslovakia, 7-11 October 1991.

[5]   Final Report TIC, Vol. II, Akadémiai Kiadó, Budapest (1994) (Z. Szatmary – Main Redactor); literature survey is this report and in the thesis; V. Lelek, "Korekce koeficientů grupových rovnic reaktoru na základe měření", ÚJV Řež (1989) (text in Czech).

[6]   Vladimír Lelek, Zoltán Szatmáry, "Operational and Safety Characteristics of Reactors with Materials Having Remarkable Indeterminateness in Data", Proceedings of the 3rd International Conference on Accelerator-Driven Transmutational Technologhy and Applications, Praha, Czech Republic, 7-11 June 1999.

[7]   V. Lelek, "Uncertainty in the Basic Nuclear Data and the Precise of the Prediction of the Planned Facilities", ANS Winter Meeting, Long Beach, California, USA, 15-18 Nov. 1999.

Appendix A

THE PROBLEMS OF THE DEPENDENCE OF
ADDING THE FUEL η ON THE PARAMETERS σ

It could be shown (in the next works) that Eqs. (4), (5) and (6) could be derived as a minimum of the functional:

$$I = \phi^+M\phi + N^+AN + \eta N^+C + \omega\big(Q(\phi,N,\sigma) - P_w\big) \tag{A1}$$

$$I = \Phi^+\aleph\Phi + \omega\big(Q(\phi,N,\sigma) - P_w\big) \tag{A2}$$

Here $\phi^+M\phi$ denotes scalar products (integration over space and time), analogically with A and $\aleph$. The $\Phi^+ = (\phi^+,N^+)$ are generalised adjoint functions. The variation of I over Φ gives us an equation for Φ^+:

$$\Phi^+\frac{\partial(\aleph\Phi)}{\partial\Phi} + \omega\frac{\partial Q}{\partial\Phi} = 0 \tag{A3}$$

Some explanation is required in order to understand these scalar products and how to cause with the operator on the left side. For example, let us consider:

$$\int\int_0^T N^+\frac{\partial\delta N}{\partial t}dtdV = \int\big[N^+(T)\delta N(T) - N^+(0)\delta N(0)\big]dV - \int\int_0^T\frac{\partial N^+}{\partial t}\delta NdtdV \tag{A4}$$

If we suppose that initial conditions for N are given then $\delta N(0) = 0$ and (A4) defines the initial conditions for $N^+(T)$ which should be zero. This is the way to proceed to have only independent variances and then to put equal zero coefficient. Thorough analyses of the functional (A1) have not yet been done, but drawing on the experiences from the literature, this should proceed without great difficulty. Particularly, using the notation of (A1):

$$\frac{\partial\eta}{\partial\sigma} = \frac{N^+\dfrac{\partial\tilde{A}}{\partial\sigma}N + \phi^+\dfrac{\partial M}{\partial\sigma}\phi + \omega\dfrac{\partial Q(\Phi,\sigma)}{\partial\sigma}}{N^+C} \tag{A5}$$

A good choice for ω is P_w, because if $P_w \to 0$ we shall automatically have from Eq. (A3) a standard adjoint equation in the linear case.

Appendix B

CORRECTIONS OF PARAMETERS WITH UNCERTAINTIES USING THE EXPERIMENTAL DATA – MEASUREMENT OF THE FUEL COMPOSITION AND OTHERS

Let us solve a problem of minimisation of a functional J, depending on parameters, with respect to the measurements as it is formulated in Eq. (10). The method we want to use is generalised Lagrange multiplicators, which consist of enlarging the J about all the conditions and equations which should be fulfilled (and multiplying them with unknown multipliers). The new functional is called Lagrangian and the sense of that operation is to change a conditional extreme to an unconditional one.

$$L = J + \Psi^+ \aleph \Phi + \left(\Phi^+ \frac{\partial(\aleph\Phi)}{\partial\Phi} + \omega\frac{\partial Q}{\partial\Phi} \right)\Psi + \omega\left(Q(\phi, N, \sigma) - P_w\right) \tag{B1}$$

and calculating $\delta L/\delta\Psi^+$, $\delta L/\delta\omega$, $\delta L/\delta\Psi$ we have our basic equation (1), (2), (3) and (A3). Varying L over Φ, Φ^+, which are also in the functional J, we shall have new systems for Ψ, Ψ^+, which could be symbolically written in the following way:

$$\Psi^+ \Im_\Phi = -\frac{\partial J}{\partial\Phi} \tag{B2}$$

$$\Im_{\Phi^+}\Psi = -\frac{\partial J}{\partial\Phi^+} \tag{B3}$$

Eqs. (B2) and (B3) are linear and we hope that can be solved without problems, as was the case for the interpretation of the measurement from the critical assembly. Relatively small problems arise, because the Ψ, Ψ^+ are not positive functions due to the quickly changing sign of their right hand side. It could be easy overcome if we do not demand a solution with a too-high level of precision.

As the result we can formulate basic steps of the numerical solution and iterative procedure for minimisation:

1. Compose functional of the parameters, which should be made more accurate.

2. Compose functional of the measurements, which should be used as an additional information about the parameters.

3. Calculate adjoint function and compose Lagrangian of the measurements and conditions, which should be respected during the minimisation.

4. Calculate Lagrange multiplier and Lagrangian.

5. Calculate via gradient method correction $\Delta\sigma$ to the parameters σ.

$$\Delta\sigma = -\varepsilon\frac{\partial L}{\partial\sigma} \tag{B4}$$

It could be shown that the minimisation procedure, represented in each iteration through the formulae (B4) converges to the minimum, if $\varepsilon > 0$ is sufficiently small.

6. We return to the Step 1 either with $\sigma_{new} = \sigma + \Delta\sigma$ if the functional is decreasing or with $\varepsilon_{new} = \varepsilon/2$ and with the same σ.

Our practical experience showed that about ten iterations are sufficient to reach the minimum. The value of ε should represent that relative correction of various parameters are expected to be roughly the same.

Concerning the time for the calculation functions Φ, Φ^+, Ψ^+, Ψ it was confirmed that it is roughly the same, only precise should be chosen slightly smaller otherwise we have oscillations. All other calculations require negligible time. From this it follows that one iteration needs about four times more time than basic calculation and that in the time about 40 times greater than the basic calculation we can find minimum in the space of our variables. For contemporary computer techniques, this is no problem.

Annex 1

LIST OF PARTICIPANTS

BELGIUM

Dr. L.H. BAETSLE
Belgian Nuclear Research Centre – SCK•CEN
Boeretang 200
B-2400 MOL

Tel: +32 14 33 22 75
Fax: +32 14 32 19 25
Eml: lbaetsle@sckcen.be

Mr. Hubert BAIRIOT
Nuclear Fuel Experts sa (FEX)
Lijsterdreef 24
B-2400 MOL

Tel: +32 (14) 31 25 33
Fax: +32 (14) 32 09 52
Eml: bairiot.fex@pophost.eunet.be

Prof. Paul A. DEJONGHE
SCK•CEN
Boeretang 200
B-2400 MOL

Tel: +32 (0)14 332595
Fax: +32 (0)14 318936
Eml: pdejongh@sckcen.be

CZECH REPUBLIC

Eng. Pavel HOSNEDL
SKODA Nuclear Machinery Ltd.
Orlik 266
CZ-316 06 PLZEN

Tel: +420 (19) 704 29 48
Fax: 48 +420 (19) 704 27 49
Eml: phosnedl@jad.ln.skoda.cz

Dr. Miloslav HRON
Nuclear Research Institute Řež, plc
CZ-250 68 REZ

Tel: +420 (0)2 66172370
Fax: +420 (0)2 20940156
Eml: hron@nri.cz

Dr. Vladimir LELEK
Nuclear Research Institute, Řež, plc
CZ-250 68 REZ

Tel: +420 2 6617 2396
Fax: +420 2 2094 0156
Eml: lelek@nri.cz

Mr. Oldrich MATAL
Energovyzkum, Ltd
Bozetechova 17
CZ-612 00 BRNO

Tel: +420 5 412 14660
Fax: +420 5 412 14659
Eml: energovyzkum@telecom.cz

Dr. Jan UHLIR
Nuclear Research Institute Řež
CZ-250 68 REZ

Tel: +420 2 6617 3548
Fax: +420 2 209 405 52
Eml: uhl@ujv.cz

FRANCE

Dr. Michel ALLIBERT
INP Grenoble
ENSEEG
BP75
F-38402 ST.-MARTIN-D'HERES

Tel: +33 (0)4 7682 6621
Fax: +33 (0)4 7682 6620
Eml: malliber@ltpcm.inpg.fr

Mr. Bertrand BARRE
Directeur R&D
COGEMA
2, rue Paul Dautier
B.P. 4
F-78141 VELIZY-VILLACOUBLAY

Tel: +33 01 39 26 30 70/3
Fax: +33 01 39 26 27 13
Eml: bbarre@cogema.fr

Ms. Catherine BESSADA
CNRS-CRMHT
1D, avenue de la recherche
F-45071 ORLÉANS Cedex 2

Tel: +33 2 3825 5509
Fax: +33 2 3863 8103
Eml: bessada@cnrs-orleans.fr

Mr. Bernard BOULLIS
CEA/VELRHO/MARCOULE
DCC/DIR – Bât. 222
B.P. 171
F-30207 BAGNOLS-SUR-CÈZE Cedex

Tel: +33(0)4 6679 6982
Fax:
Eml: bernard.boullis@cea.fr

Mr. Gilles BOURGES
CEA VALDUC
BP 120
F-21120 IS-SUR-TILLE

Tel: +33 03 80 23 43 97
Fax: +33 03 80 23 52 19
Eml:

Mr. Hubert BOUSSIER
Commissariat à l'Énergie Atomique
CEN Valrho Marcoule
B.P. 171
F-30207 BAGNOLS-SUR-CÈZE

Tel: +33 66 79 63 28
Fax: +33 66 79 63 48
Eml: hubert.boussier@cea.fr

Mr. P. BROSSARD
DCC/DRDD/SPHA/SAED
CEN de Marcoule
BP.171
F-30207 BAGNOLS-SUR-CÈZE Cedex

Tel: +33 66 79 65 64
Fax: +33 66 79 65 67
Eml: brossardp@amandine.cea.fr

Mr. Bernard CARLUEC
FRAMATOME
10, rue Juliette Récamier
F-69456 LYON Cedex 06

Tel: +33 4 72 74 70 66
Fax: +33 4 72 74 73 30
Eml: bcarluec@framatome.fr

Mr. Luc CHAUDON
CEA Valrho
BP 171
F-30207 BAGNOLS-SUR-CÈZE

Tel: +33 4 6679 6609
Fax: +33 4 6679 6617
Eml: luc.chaudon@cea.fr

Mr. Olivier CONOCAR
CEA VALRHO
SPHA-LEPP-BP171
F-30207 BAGNOLS-SUR-CÈZE

Tel: +33 4 6679 6311
Fax: +33 4 6679 6567
Eml: conocar@amandine.cea.fr

Mr. Didier DEVILLARD
CEA Valduc
BP 120
F-21120 IS-SUR-TILLE

Tel: +33 3 8023 4921
Fax: +33 3 8023 5219
Eml:

Mr. Hubert DOUBRE
Directeur, CSNSM
IN2P3-CNRS
Bât. 104
F-91405 ORSAY Cedex

Tel: +33 1 69 41 52 30
Fax: +33 1 69 41 50 08
Eml: doubre@csnsm.in2p3.fr

Prof. Augusto GANDINI
CEA-DRN
CEN Cadarache
F-13108 ST.-PAUL-LEZ-DURANCE

Tel: +33 4 4225 2257
Fax: +33 4 4225 4242
Eml: gndna@tin.it

Prof. Marcelle GAUNE-ESCARD
IUSTI
Technopole de Château-Gombert
5 rue Enrico Fermi
F-13453 MARSEILLE Cedex 13

Tel: +33 4 91 10 68 87
Fax: +33 4 91 11 74 39
Eml: mge@iusti.univ-mrs.fr

Dr. Konstantin GUERMAN
Director of Researches
UMR 5084/CNRS
Le Haut Vigneau
F-33175 GRADIGNAN

Tel: +33 620712426/557 12
Fax: +33 5 5612 0900
Eml: guerman@cenbg.in2p3.fr

Prof. Robert GUILLAUMONT
Commission Nationale d'Évaluation
Tour Mirabeau, 39-43 Quai André-Citroen
F-75015 PARIS

Tel: +33 1 4058 8905
Fax: +331 4058 8938
Eml: C.Jouvance.cne@wanadoo.fr

Mr. Hervé HANCOK
CEA Valrhô
F-30207 BAGNOLS-SUR-CÈZE

Tel: +33 4 6679 6247
Fax: +33 4 6679 6348
Eml: herve.hancok@cea.fr

Prof. Sergey A. KUZNETSOV
IUSTI
Technopole Château-Gombert, 5
5 rue Enrico Fermi
F-13453 MARSEILLE Cedex 13

Tel: +33 4 9110 6881
Fax: +33 4 9111 7439
Eml: kuznet@iusti.univ-mrs.fr

Mr. Jérôme LACQUEMENT
CEA Marcoule
B.P. 171
F-30207 BAGNOLS-SUR-CÈZE

Tel:
Fax:
Eml: lacquement@amandine.cea.fr

Mr. David LAMBERTIN
CE Valrhô (DCC/DRRV/SPHA/LEPP)
BP 371
F-30207 BAGNOLS-SUR-CÈZE

Tel: +33 4 66 79 63 11
Fax: +33 4 66 79 65 67
Eml: lambertind@amandine.cea.fr

Dr. Frederic LANTELME
Director, CNRS
LI2C-Électrochimie, case51,
4 Place Jussieu
F-75252 PARIS Cedex 05

Tel: +33 1 4427 3191
Fax: +33 1 4427 3834
Eml: frl@ccr.jussieu.fr

Ms. Annabelle LAPLACE
DCC/DRRV/SPHA
Bât. 166
CEA Valrhô
F-30207 BAGNOLS-SUR-CÈZE

Tel: +33 (0)4 66 79 65 49
Fax: +33 (0)4 66 79 65 67
Eml: laplacea@amandine.cea.fr

Dr. Michel LECOMTE
Direction Technique
FRAMATOME
Tour Fiat, Cedex 16
F-92084 PARIS-LA-DÉFENSE

Tel: +33 0(1) 47 96 56 73
Fax: +33 0(1) 47 96 15 09
Eml: mlecomte@framatome.fr

Prof. Klans LUTZENKIRCHEN
IRes, Chimie Nucléaire
Univ. Louis Pasteur
23 Rue Du Loess
F-67037 STRASBOURG CX2

Tel: +33 03 88 10 64 04
Fax: +33 03 88 10 64 31
Eml: klutz@in2p3.fr

Prof. Charles MADIC
CEA/Saclay
Direction du Cycle du DCC/Dir. Bt 450
F-91191 GIF-SUR-YVETTE

Tel: +33 1 69 08 82 07
Fax: +33 1 69 08 85 38
Eml: charles.madic@cea.fr

Dr. Hervé NIFENECKER
Institut des Sciences Nucléaires
53, ave. des Martyrs
F-38026 GRENOBLE Cedex

Tel: +33 (0)4 76 28 40 62
Fax: +33 (0)4 76 28 40 04
Eml: nif@in2p3.fr

Mr. Jean-Christophe PETIT
Project Engineer
SGN
1, rue de Hérons
F-78182 MONTIGUY-LE-BX

Tel: +33 1 3948 7412
Fax: +33 1 3948 5990
Eml: jch.petit@wanadoo.fr

Dr. Gérard PICARD
CNRS UMR 7575
École Nationale Supérieure de Chimie de Paris
11 rue P. et M. Curie
F-75231 PARIS Cedex 05

Tel: +33 1 55 42 63 89
Fax: +33 1 44 27 67 50
Eml: picard@ext.jussieu.fr

Mme Sylvie PILLON
CEA-DRN/DER/SIS
Centre d'Études de Cadarache
F-13108 ST.-PAUL-LEZ-DURANCE

Tel: +33 (0)4 42 25 40 67
Fax: +33 (0)4 42 25 40 46
Eml: Sylvie.Pillon@cea.fr

Ms. Catherine RABBE
CEA Marcoule
B.P. 171
F-30207 BAGNOLS-SUR-CÈZE

Tel:
Fax:
Eml: rabbe@amandine.cea.fr

Mr. Tobias REICH
ROBL
CRG
ESRF
B.P. 220
F-38043 GENOBLE Cedex

Tel: +33 (0)4 76 88 23 39
Fax: +33 (0)4 76 88 25 05
Eml: reich@esrf.fr

Prof. Massimo SALVATORES
Directeur de Recherche
Dir. des Réacteurs Nucléaires
Bat. 707 – CE Cadarache
F-13108 ST.-PAUL-LEZ-DURANCE Cedex

Tel: +33 (0)4 42 25 33 65
Fax: +33 (0)4 42 25 41 42
Eml: massimo.salvatores@cea.fr

Dr. Sylvie SANCHEZ
CNRS
Labo Électrochimie et Chimie
ENSCP
F-75005 PARIS

Tel: +33 1 5542 6391
Fax: +33 1 4427 6750
Eml: sanchez@ext.jussieu.fr

Mr. Bruno SICARD
Assistant du Directeur
CEA/DCC/DIR
CEN Marcoule – B.P. 1
F-30200 BAGNOLS-SUR-CÈZE

Tel: +33 (0)4 66 79 69 44
Fax: +33 (0)4 66 79 66 53
Eml: sicard@amandin.cea.fr

Dr. Monique SIMONOFF
Director
UMR 5084/CNRS
Domaine Le Haut Vigneau
F-33175 GRADIGNAN

Tel: +33 5 5712 0902
Fax: +33 5 5712 0900
Eml: simonoff@cenbg.in2p3.fr

Mr. Christian SOREL
CEA Valrhô
DCC/DRRV/SEMP/LMP
Bâtiment ATALANTE BP 171
F-30207 BAGNOLS-SUR-CÈZE

Tel: +33 4 6679 1639
Fax: +33 4 6679 1630
Eml: christian.sorel@cea.fr

Mr. Pierre TAXIL

Tel:
Fax:
Eml: taxil@ramses.ups-tlse.fr

INDIA

Mr. Krishnamruthy NAGARAJAN
Head, Pyrochemical Processing Section
Fuel Chemistry Division
Indira Gandhi Center for Atomic Research
Kalpakkam 603 102
TAMIL NADU

Tel: +91 04114 40398
Fax: +91 04114 40365
Eml: knag@igcar.ernet.in

Mr. B. Prabhakara REDDY
Indira Gandhi Center for Atomic Research
Kalpakkam, 603 102
TAMIL NADU

Tel: +91 4114 40398
Fax: +91 4114 40365
Eml: vasu@igcar.ernet.in

Dr. P.R. VASUDEVA RAO
Head, Fuel Chemistry Division
Indira Gandhi Center for Atomic Research
Kalpakkam 603 102
TAMIL NADU

Tel:
Fax:
Eml: vasu@igcar.ernet.in

ITALY

Dr. Giuseppe MARUCCI
ENEA/CASACCIA
ERG
S.P.054, Via Anguillarese
00100 ROMA

Tel: +39 (6) 3048 3106
Fax: +39 (6) 3048 6308
Eml: marucci@casaccia.enea.it

JAPAN

Dr. Tadashi INOUE
Director, Nuclear Fuel Cycle Dept.
Komae Research Lab., CRIEPI
2-11-1 Iwato-kita
Komae-shi, TOKYO 201

Tel: +81 3 34 80 21 11
Fax: +81 3 34 80 79 56
Eml: inouet@criepi.denken.or.jp

Mr. Shinichi KITAWAKI
Japan Nuclear Cycle
4-33, Muramatsu
Tokai-mura, Naka-gun, IBARAKI
319-1194

Tel: +81 29-282-1111
Fax: +81 29-287-0685
Eml: sin@tokai.jnc.go.jp

Dr. Tadafumi KOYAMA
Senior Researcher
CRIEPI
2-11-1 Iwadokita, Komae-shi
TOKYO 201-0004

Tel: +81 +49 7247 951 490
Fax: +81 +49 7247 951 593
Eml: koyama@criepi.denken.or.jp

Prof. Hirotake MORIYAMA
Kyoto University
Kumatori-cho, Sennan-gun
590-0494

Tel: +81 724 51 2424
Fax: +81 724 51 2634
Eml: moriyama@rri.kyoto-u.ac.jp

Dr. Toru OGAWA
Group Leader
Research Group for Actinides
JAERI
Tokai-Mura, Naka-gun
IBARAKI-KEN 319-1195

Tel: +81 29 282 5382
Fax: +81 29 282 5922
Eml: ogawa@molten.tokai.jaeri.go.jp

Dr. Hajimu YAMANA
Research Reactor Institute
Kyoto University
Noda, Kumatori-cho, Sennan-gun
590-0494

Tel: +81 724 51 2442
Fax: +81 724 51 2634
Eml: yamana@HL.rri.kyoto-u.ac.jp

KOREA (REPUBLIC OF)

Dr. Joon-Bo SHIM
KAERI
P.O. Box 105, YUSUNG
Taejun 305-600

Tel: +82 42 868 2340
Fax: +82 42 868 2329
Eml: njbshim@nanum.kaeri.re.kr

THE NETHERLANDS

Dr. Mark HUNTELAAR
NRG
Westerduinweg 3
PO Box 25
1755 ZG PETTEN

Tel: +31 224 56 4042
Fax: +31 224 56 1883
Eml: huntelaar@nrg-nl.com

RUSSIAN FEDERATION

Dr. Alexandre V. BYCHKOV
SSC Research Institute of
DIMITROVGRAD-10
433510

Tel: +7 84235 32021
Fax: +7 84235 35648/65554
Eml: bav@niiar.simbirsk.su

Mr. Victor V. IGNATIEV
RRC – Kurchatov Institute
I.V. Kurchatov Sq. 1
123182 MOSCOW

Tel: +7 95 196 71 30
Fax: +7 95 196 86 79
Eml: ignatiev@quest.net.kiae.su

Mr. Michael V. KORMILITZYN
Head of Fuel Technology
SSC Research Institute of
SSC RIAR
DIMITROVGRAD-10
433510

Tel: +7 84235 32021
Fax: +7 84235 35648
Eml: bav@niiar.simbirsk.su

Prof. Sergey A. KUZNETSOV
Institute of Chemistry Kola
14 Fersman St.
184200 APATITY

Tel: +7 815 55 79 730
Fax: +7 47 789 14 131
Eml: kuznet@chemy.kolasc.net.ru

SPAIN

Dr. Concepcion CARAVACA
CIEMAT
Avda. Complutense 22
E-28040 MADRID

Tel: +34 91 34 662 16
Fax: +34 91 34 662 33
Eml: c.caravaca@ciemat.es

Ms. Paloma DIAZ AROCAS
Inst. Tecnologia Nuclear
CIEMAT
Avda. Complutense 22
E-28040 MADRID

Tel: +34 91 346 62 90
Fax: +34 91 346 62 33
Eml: p.diazarocas@ciemat.es

UNITED KINGDOM

Dr. Richard BUSH
AEA Technology
220 Harwell
DIDCOT
Oxfordshire OX11 ORA

Tel: +44 1235 43 49 12
Fax: +44 1235 43 61 33
Eml: richare.bush@aeat.co.uk

Mr. David HEBDITCH
BNFL
Berkeley Centre, BERKELEY
Gloucestershire
GL13 9PB

Tel: +44 01453 813156
Fax: +44 01453 813335
Eml: hebditch@tokai.jnc.go.jp

Mr. Jon JENKINS
AEA Technology – Nuclear
220 Harwell
DIDCOT, Oxfordshire
OX11 0RA

Tel: +44 1235 434878
Fax: +44 1236 436133
Eml: jon.jenkins@aeat.co.uk

Mr. Robert THIED
BNFL
B229, Sellafield, SEASCALE
CA20 1PG

Tel: +44 019467 74594
Fax: +44 019467 85740
Eml: rob.c.thied@bnfl.com

Dr. Peter David WILSON
BNFL Sellafield
B229, Sellafield
SEASCALE, Cumbria CA 20 IPG

Tel: +44 19 467 75168
Fax: +44 19 467 76984
Eml: pdw1@bnfl.com

UNITED STATES OF AMERICA

Prof. Gregory CHOPPIN
Dept. of Chemistry, B-164
Florida State University
TALLAHASSEE, Florida 32306-4390

Tel: +1 (904) 644 3875
Fax: +1 (904) 644 8281
Eml: choppin@chemmail.chem.fsu.edu

Dr. James J. LAIDLER
Argonne National Laboratory
Chemical Technology Division
9700 South Cass Avenue, Bldg
ARGONNE, Illinois 60439-4837

Tel: +1 630 252 4479
Fax: +1 630 972 4479
Eml: laidler@cmt.anl.gov

Dr. Wayne A. PUNJAK
Los Alamos National Laboratory
P.O. Box 1663
NMT-2, Mail Stop E511
LOS ALAMOS, New Mexico 87545

Tel: +1 505 667 4630
Fax: +1 505 665 1780
Eml: wpunjak@lanl.gov

Dr. Mac TOTH
170 Bayview Drive
TEN MILS, TN 37880

Tel: +1 865 376 3753
Fax: +1 865 717 9788
Eml: marimac@icx.net

INTERNATIONAL ORGANISATIONS

Dr. Jae Sol LEE
IAEA
Wagramerstrasse 5
A-1400 VIENNA

Tel: +43 1-2600-22767
Fax: +43 1-26007
Eml: J.S.Lee@iaea.org

Dr. Michel HUGON
DG XII/F-5, MO75 5/55
European Commission
200, rue de la Loi
B-1049 BRUXELLES

Tel: +32 (0)2 296 57 19
Fax: +32 (0)2 295 49 91
Eml: Michel.Hugon@cec.eu.int

Dr. Claes NORDBORG
OECD/NEA Data Bank
Le Seine Saint Germain
12, boulevard des Iles
F-92130 ISSY-LES-MOULINEAUX

Tel: +33 (1) 4524 1090
Fax: +33 (1) 4524 1110
Eml: nordborg@nea.fr

Mr. Satoshi SAKURAI
Nuclear Science Division
OECD Nuclear Energy Agency
Le Seine Saint Germain
12, boulevard des Iles
F-92130 ISSY-LES-MOULINEAUX

Tel: +33 01 4524 1152
Fax: +33 01 4524 1106
Eml: sakurai@nea.fr

Mr. Philippe SAVELLI
Deputy Director
OECD Nuclear Energy Agency
Le Seine Saint Germain
12, boulevard des Iles
F-92130 ISSY-LES-MOULINEAUX

Tel: +33 (1) 45 24 10 06
Fax: +33 (1) 45 24 11 10
Eml: philippe.savelli@oecd.org

Dr. Lothar KOCH
Head of Nuclear Chemistry
Institute for Transuranium Elements
Postfach 2340
D-76125 KARLSRUHE

Tel: +49 7247 951 424
Fax: +49 7247 951 596
Eml: koch@itu.fzk.de

Dr. Tadafumi KOYAMA
European Commission
Joint Research Centre
Institute for Transuranium
Postfach 2340
D-76125 KARLSRUHE

Tel: +49.7247.951490
Fax: +49.7247.951593
Eml: koyama@itu.fzk.de

Ms. Carole PERNEL
ITU
Postfach 2340
D-76125 KARLSRUHE

Tel: +49 7247 951 281
Fax: +49 7247 951 561
Eml: carole.pernel@itu.fzk.de

Dr. YACINE KADI
Emerging Energy Technologies
SL Division
European Organisation for Nuclear Research
CH-1211 GENEVE 23

Tel: +41 22 7679569
Fax: +41 22 7677555
Eml: Yacine.Kadi@cern.ch

Annex 2

WORKSHOP ORGANISATION

Chairperson	
C. Madic	CEA, France

International Scientific Advisors	
B. Boullis	CEA, France
H. Boussier	CEA, France
Ph. Brossard	CEA, France
G. Choppin	Florida State University, USA
M. Gaune-Escard	University of Provence, France
M. Hugon	EC, Belgium
T. Inoue	CRIEPI, Japan
L. Kock	ITU, Germany
C. Nordborg	OECD/NEA, France
T. Ogawa	JAERI, Japan
M. Salvatores	CEA, France
J. Uhlir	NRI, Czech Republic
P. Wilson	BNFL, UK

Local Organising Committee	
Ph. Brossard	CEA, France (20/01/2000)
A. Laplace	CEA, France (11/02/2000)
I. Olivier	CEA, France (11/02/2000)

Workshop Secretariat	
C. Faux	CEA, France
S. Sakurai	OECD/NEA France

Annex 3

STATE-OF-THE-ART REPORT ON PYROCHEMICAL SEPARATIONS*

Background

In the future, nuclear fuel will reach higher burn-up and recycling of spent fuel will have to be performed after shorter cooling times to be economically and ecologically attractive. Dry reprocessing using pyrochemical methods received for a number of decades the attention of some research institutes, for example the RIAR (Dimitrovgrad) in Russia, ANL (Illinois, Idaho) in the USA, CRIEPI in Japan, and institutes in France and Belgium.

Pyrochemical separation can be classified into several techniques: volatilisation, liquid-liquid extraction using either immiscible molten metal phases or immiscible molten metal-molten salt phases, electrorefining in non-aqueous media, vacuum distillation, fractional crystallisation, melt refining, zone melting and gas-solid reactions. These methods aim at the separation of minor actinides, as well as the recovery of useful elements. The proposed processes are very complex and necessitate the use of highly controlled atmospheres to avoid hydrolysis and precipitation reactions. However, these processes have an inherent advantage in criticality safety considerations, in the compactness of the plant and for the stability against the high radiation dose. Except for the pilot-scale demonstration of pyroelectrolysis at Argonne-West and RIAR, all other studies are at the laboratory scale.

Partitioning and transmutation (P&T) of long-lived fission products and minor actinides are considered in future radioactive waste management scenarios. Multi-recycling of the targets or fuel will be required in order to achieve an efficient transmutation in dedicated systems (e.g. fast reactors or accelerator-driven systems). If other fuel types, e.g. nitride fuel or metallic fuels, are to be used in multi-recycling processes, pyrochemical methods might become necessary for the implementation of this option.

Extensive R&D work will be required in order to upgrade the pyrochemical separation process to the level of present industrial aqueous reprocessing. However, in order to shorten the cooling times, which determine the recycling period of highly irradiated fuel, the development of alternative recycling processes to the present PUREX process will become mandatory.

Objectives

The study would focus on reviewing the status of relevant technologies and surveying the feasibility and potential roles of pyrochemical separations (mainly the extraction processes involving the use of molten salts) for future fuel cycle options. The study would be carried out in co-operation with other interested international organisations, such as IAEA, EC, and ISTC.

---

* This proposal was first discussed at the Tenth Meeting of the OECD/NEA Nuclear Science Committee (2-4 June 1999). The NSC decided to organise a workshop on pyrochemistry as a starting point for the collection of information to be included in a state-of-the-art report.

The study would cover:

- Review of the previous studies.

- Survey of R&D programmes.

- Identification of technical issues and challenges.

- Assessment of technical feasibility.

- Recommendation on future direction of R&D.

- Needs of international collaboration.

Working method

A task force would be set up. The first meeting of the task force would be convened in late 2000 to review major activities and international co-operation concerning pyrochemical separations. Two or three meetings may be necessary to review the R&D activities and prepare a report to be published in 2002.

ALSO AVAILABLE

NEA Publications of General Interest

1999 Annual Report (2000)   *Free: available on Web.*

NEA News
ISSN 1605-9581   Yearly subscription: FF 240 US$ 45 DM 75 £ 26 ¥ 4 800

Geologic Disposal of Radioactive Waste in Perspective (2000)
ISBN 92-64-18425-2   Price: FF 130 US$ 20 DM 39 £ 12 ¥ 2 050

Radiation in Perspective – Applications, Risks and Protection (1997)
ISBN 92-64-15483-3   Price: FF 135 US$ 27 DM 40 £ 17 ¥ 2 850

Radioactive Waste Management in Perspective (1996)
ISBN 92-64-14692-X   Price: FF 310 US$ 63 DM 89 £ 44

Nuclear Science

Core Monitoring for Commercial Reactors: Improvements in Systems and Methods (2000)
ISBN 92-64-17659-4   Price: FF 460 US$ 71 DM 137 £ 44 ¥ 7 450
Ion and Slow Positron Beam Utilisation (1999)
ISBN 92-64-17025-1   Price: FF 400 US$ 72 DM 119 £ 43 ¥ 8 500
Physics and Fuel Performance of Reactor-Based Plutonium Disposition (1999)
ISBN 92-64-17050-2   Price: FF 400 US$ 70 DM 119 £ 43 ¥ 8 200
Shielding Aspects of Accelerators, Targets and Irradiation Facilities (SATIF-4)(1999)
ISBN 92-64-17044-8   Price: FF 500 US$ 88 DM 149 £ 53 ¥ 10 300

3-D Radiation Transport Benchmarks for Simple Geometries with Void Regions
(2000)   *Free on request.*
Benchmark Calculations of Power Distribution Within Fuel Assemblies
Phase II: Comparison of Data Reduction and Power Reconstruction Methods in Production Codes
(2000)   *Free on request.*
Benchmark on the VENUS-2 MOX Core Measurements
(2000)   *Free on request.*
Calculations of Different Transmutation Concepts: An International Benchmark Exercise
(2000)   *Free on request.*
Prediction of Neutron Embrittlement in the Reactor Pressure Vessel: VENUS-1 and VENUS-3 Benchmarks
(2000)   *Free on request.*
Pressurised Water Reactor Main Steam Line Break (MSLB) Benchmark
(2000)   *Free on request.*

International Evaluation Co-operation *(Free on request)*

Volume 1: *Comparison of Evaluated Data for Chromium-58, Iron-56 and Nickel-58* (1996)
Volume 2: *Generation of Covariance Files for Iron-56 and Natural Iron* (1996)
Volume 3: *Actinide Data in the Thermal Energy Range* (1996)
Volume 4: *^{238}U Capture and Inelastic Cross-Sections* (1999)
Volume 5: *Plutonium-239 Fission Cross-Section between 1 and 100 keV* (1996)
Volume 8: *Present Status of Minor Actinide Data* (1999)
Volume 12: *Nuclear Model to 200 MeV for High-Energy Data Evaluations* (1998)
Volume 13: *Intermediate Energy Data* (1998)
Volume 14: *Processing and Validation of Intermediate Energy Evaluated Data Files* (2000)
Volume 15: *Cross-Section Fluctuations and Shelf-Shielding Effects in the Unresolved Resonance Region* (1996)
Volume16: *Effects of Shape Differences in the Level Densities of Three Formalisms on Calculated Cross-Sections* (1998)
Volume 17: *Status of Pseudo-Fission Product Cross-Sections for Fast Reactors* (1998)
Volume 18: *Epithermal Capture Cross-Section of ^{235}U* (1999)

Order form on reverse side.

ORDER FORM

OECD Nuclear Energy Agency, 12 boulevard des Iles, F-92130 Issy-les-Moulineaux, France
Tel. 33 (0)1 45 24 10 10, Fax 33 (0)1 45 24 11 10, E-mail: nea@nea.fr, Internet: www.nea.fr

Qty	Title	ISBN	Price	Amount
			Postage fees*	
			Total	

*European Union: FF 15 – Other countries: FF 20

❏ Payment enclosed (cheque or money order payable to OECD Publications).

Charge my credit card   ❏ VISA   ❏ Mastercard   ❏ Eurocard   ❏ American Express

(N.B.: You will be charged in French francs).

Card No.	Expiration date	Signature
Name		
Address	Country	
Telephone	Fax	
E-mail		